BEHAVIOURAL PROBLEMS OF FARM ANIMALS

BEHAVIOURAL PROBLEMS OF FARM ANIMALS

by

M. Kiley-Worthington

ORIEL PRESS

STOCKSFIELD

LONDON   HENLEY-on-THAMES   BOSTON

First published in 1977
by Oriel Press Ltd (Routledge & Kegan Paul Ltd)
Branch End, Stocksfield, Northumberland NE43 7NA

Printed and bound in England by
Knight and Forster Ltd, Leeds

ACKNOWLEDGEMENTS

The project resulting in this monograph was financed by the Agricultural Research Council, Britain. The RSPCA are thanked for financing some of the time for revision and publication.

Many farmers and research workers gave up their time to discuss aspects of this work with me. I am particularly indebted to the partners of Colin Godmans Farm, Sussex.

Several people have critically read all or part of this monograph. I would like to thank Professor Andrew for allowing me to work in his department and for interest and encouragement as well as critical comment. Dr D. Wood-Gush's constructive criticism was also particularly appreciated. In addition, Drs J. Archer, D. Clayton and J. Davey are thanked for critically reading parts of the manuscript.

The Commonwealth Foundation are thanked for financing my study tour to Canada which added considerable depth to the work.

Finally, the arduous tasks of typing and checking bibliography were cheerfully dealt with by Barbara Garratt, Jill Illot and Beryl Blackburn, to whom I am very grateful.

Chapters 10 and 15 have been published elsewhere and are reproduced here with the permission of the editor of the British Veterinary Journal and the International Union for the Conservation of Nature respectively.

CONTENTS

1

INTRODUCTION

For the past half century, animal husbandry research has encompassed nutrition, animal breeding and physical pathologies. In so far as environmental design has been considered, the emphasis has been on labour-saving devices and hygiene. The normal behaviour of the animal and the effect of different environments on this behaviour has received little attention. The results of such an approach are that only too often "the buildings are designed for the comfort of the caretaker, rather than the pigs (or other stock)." (Ross 1960). There are some exceptions, however, to this approach , and these have been among the husbandry men who keep their animals largely for behavioural performance (such as horse training and riding establishments, some circuses and occasional zoos). Here the stockmen have been aware that inadequate environments, either in the social or physical sense, will affect the animals' training. Hediger (1955) was the first to point out some of the ways in which a knowledge of behaviour could lead to better environmental design from the animals' point of view. Further knowledge, particularly in the agricultural field, has been regrettably slow in its accumulation.

It is hardly necessary at this stage to outline the reasons for an interest in behavioural problems in animal husbandry but perhaps these should be stated. Firstly, there are economic reasons. Because profit margins in relation to capital investment are usually very small on the farm, any loss of these as a result of a lack of environmental design and husbandry techniques which do not take into account the animals' psychological as well as physiological demands should be avoided. This factor will, of course, become increasingly important with intensification of animal husbandry. It will also become increasingly important as the price of animal food stuffs increases, forcing meat prices up. Animal production, compared with protein production directly from plants, is relatively inefficient. However, in Britain today, for example, it still absorbs the lion's share of available energy (Blaxter 1974). Unless efficiency is increased, growing world populations will ensure that most animal proteins become a luxury.

The second reason why a consideration of behavioural effects and problems of animal husbandry is important is that it has been suggested that such intensification of animal husbandry necessitates a decline in animal welfare. Questions such as 'what is cruel?' and 'what is not?' for individual animals or species are questions that cannot possibly be answered without much more behavioural research, and a real comprehension of the social and physical environmental demands of the animal, as well as serious consideration of ethical aspects.

There is no clash of interests between the animal productionist and welfarist since sub-optimal environments for the animal will benefit neither interest group. Thus an understanding of behaviour, normal and abnormal, and the limits of adaptability of behaviour in different species, age groups and individuals is the only way to achieve optimal production and welfare.

A second reason for this monograph was to investigate how behaviour can be influenced by environmental design. This encompasses a multi-disciplinary approach since a satisfactory understanding of this (and solutions to any behavioural problems) cannot be found until their cause is understood. Thus it is imperative to consider current ideas on such problems from the psychological, ethological and ecological view point, as well as the more pragmatic approach of the agriculturalist and veterinarian. There is in addition a wealth of knowledge in the head and hands of skilled animal husbandrymen. During the course of this study interviews and farm visits were conducted in many farms, zoos, wildlife parks and research institutes. This had value both in the identification of the behavioural problems faced by the husbandrymen and also in pointing out the practical problems in suggesting environmental improvements.

One of the striking things is the astonishing lack of basic etho-ecological knowledge on the domestic ungulates in particular. The situation is so absurd that in many respects we know more today about the etho-ecology of the American big horn sheep, or the Soay sheep of St Kilda (e.g. Geist 1971 and Jewell et al. 1974), than we do about the behaviour and ecology of the domestic sheep! It is essential to have information on social organization, habitat preferences and so on before we know what type of environment the animals have evolved to be physically and psychologically adapted to. Thereafter an understanding of such aspects can be used to improve the farm environment. The criticism has been made that, as a result of domestication, domestic animals have changed so much that they no longer resemble their wild contemporaries in behaviour and social organization. Existing studies of feral populations indicate that the social organization of the domestic animal remains potentially very complex and has changed little as a result of domestication. Thus Grubb & Jewell (1966) studied the Soay sheep on St Kilda and found their social organization to be very similar to that of domestic sheep described by Scott (1945). The social behaviour of the wild boar has been studied by Gundlack (1968). The similarity in social organization and habitat choice of what we know of the feral domestic pig is striking (e.g. Hanson & Lars 1959) despite their obvious morphological changes. In addition, such details as their vocalizations and the situations that elicit them are virtually indistinguishable (Kiley 1972). Even dogs, one of the animals that has been domesticated longest and is kept in environments very different from that for which they evolved, revert to pack marauding behaviour similar to that of other social canids (e.g. the wolf and wild dog, Mech 1962 and Van Lawick-Goodall 1970) when given the opportunity, for example at the end of the second World War in France.

A summary of what is known of the social organization, sociability, habitat preferences and communication systems of groups of feral domestic animals or, in their absence, in their wild close relatives is given in Table 1. It is apparent even with such broad categories of behavious that there is little known on these groups and, of course, we need far more detailed studies. Even this amount of information indicates to the husbandryman the fundamental size and composition of groups of each species. Clearly, the more social animals should be able to live in larger groups without showing any abnormal behavioural or physiological changes. Thus cattle can be kept successfully in larger herds than the less social pig. Conversely, the effects of isolation from other conspecifics is likely to be more traumatic for the social than for the solitary animal.

In this monograph the behavioural problems are identified. They are then treated individually, each one being the subject of a short review which attempts an assessment of both its cause and effect. Possible solutions to these problems are discussed with each topic and areas particularly in need of proper behavioural research are pointed out. Finally there is a short section on housing.

CHAPTER 1
AN EXAMINATION OF THE CONCEPT OF STRESS

It is impossible to avoid a discussion of stress in a monograph of this type. The term first arrived in the physiological literature when Selye (1950) defined it as a general condition (consisting of several physiological and neurophysiological changes in the organism) as a result of certain strong external or internal stimuli, described as 'stressors'. When the organism is under the influence of such stressors (those particularly mentioned by Selye were heat, cold, pain and bacterial infection) then he suggested that the organism reacted in an appropriate way to combat such stimuli and reduce their destructive effect. The syndrome of physiological change that results he termed the General Adaptive Syndrome (G.A.S.). It must be emphasised, as Selye points out, that the effect of such physiological changes (which involve principally an increase in autonomic nervous activity, changes in blood corticosteroids and ACTH levels) in the first instance was to increase the animal's survival. They were, therefore, necessary, desirable responses. This was well demonstrated by Ferguson et al. (1970) who showed that 'stressed' mice (in this work social-stress is used, that is, the introduction of strange individuals into a cage) survived better when *Trypnosoma brucii* was injected than the isolates did. Selye quotes many other examples of the G.A.S. aiding survival.

On the other hand, it is clear that prolonged activation of the G.A.S. has harmful effects on the animals and survival is reduced as a result. Van der Welt & Jansen (1968) give a good summary of the various stages of the G.A.S. alarm reaction consisting of shock and counter shock, followed by the stage of resistance and finally exhaustion. They list the physiological and neuro-hormonal changes that have been shown to take place at each of these stages.

As soon as the neuro-hormonal link with the General Adaptive Syndrome was recognised, behavioural and psychological workers began to latch on to the concept of 'stress' and use it as a descriptive and explanatory term to describe an environment in which the animal was under adverse psychological pressure. Under these conditions either it was assumed that similar physiological changes were taking place as when the environment was 'stressful' and, indeed, in some cases corticosteroid levels were even measured (Brady 1964), or no reference was made to the physiological usage of the term, while it remained behaviourally undefined (e.g. McBride 1968).

Further, the word 'stress' shortly became an explanatory term; any disease or behavioural malfunction that could not be explained in any other way was noted as the result of 'stress'. Indeed, it has been defined as 'any adverse stimulus for the animal' (Coffey 1971). Such statements make no contribution to reducing the confusion surrounding this concept since they say nothing whatever until we know what we can describe as adverse to what animal under what conditions, and measure it.

McBride (1968) made an attempt at a behavioural measurement of 'social stress'. He evolved a formula to measure social stress in which he begins with the symbol 'R'. This is defined as 'the amount of space the animal requires, being the amount it occupies physically plus *the amount it requires to control the intensity of social stimuli*'.

The mathematics appear straightforward, but such a formula is quite unhelpful since the problem is to assess '*the amount (of space) it requires to control the intensity of social stimuli*'. It is not clear what is meant by 'intensity of social stimuli' and how this could be measured. Since McBride is not relating 'social stress' to physiological stress in the way the term was originally coined, such an approach only adds to the confusion.

The physiological responses to stress are by no means simple or constant. They appear to depend principally on the length of the stressing stimulus, as well as its intensity, and the past experience of the animal (Ader & Friedman 1964). Most systems of the body are involved to varying degrees. Thus it has been shown that resistance to disease is increased if a degree of stress is administered first, whether this administered stress is physical (Freeman 1971) or social (Gross & Siegel 1965; Ferguson 1969; Ferguson et al. 1970).

In this context, the incidence of herd disease should be looked at. This is a much neglected approach to veterinary medicine. Epidemics and diseases in intensive husbandry continue to increase, yet the present day solution is:-

(1) To increase the level of hygiene, usually by increasing disinfectant measures.

(2) To feed base levels of antibiotics throughout life.

(3) To isolate animals, ostensibly to avoid transfer of the disease. The effect of isolation may well be to increase the susceptibility to disease by changes in the General Adaptive Syndrome.

Ader & Friedman reviewed the literature in 1964 and showed there was experimental evidence that housing, early experience, isolation from mother and other factors will affect the susceptibility to disease. They concluded that "pathogenic stimuli do not act in a vacuum, but are superimposed on a background of experimentally determined psycho-physiological reactivity. It is the interaction between what already exists within the organism and responses induced by specific pathogenic stimuli that will determine relative susceptibility in differentially manipulated animals."

With this in mind it is suggested that it would be of considerable use to look at the history of herd disease on farms. It is possible that certain management conditions which are from the animal's point of view likely to lead to physiological responses of stress could be reflected in an increased susceptibility to disease. Disease is not solely the result of chance infection: even mastitis has been shown to be correlated with reduced ascorbic acid levels (symptomatic of adreno-cortical activity) and thus is not due only to the presence of streptococci (Reineke et al. 1941).

Selye's main, and very considerable, contribution to the study of stress was that he pointed out that the physiological responses were non-specific, no matter what the type of stress was, be it heat, cold, pain or social stress of one sort or another. Indeed, the physiological responses to such conditions have since been confirmed as being non-specific to the stimulus and mainly involving the adrenal and pituitary axis. Nevertheless, work since Selye has shown that many other complex and little understood mechanisms in the body are involved (see, for example, Van der Welt & Janson's 1968 impressive list of responses).

The veterinary profession has been called upon to set up 'codes of practice' (Ministry of Agriculture, Fisheries and Food 1971). These publications abound in vague statements such as "The space allowance for cattle housed in groups should be calculated in relation to the whole environment, the age of the stock and the size of group, and should be based on appropriate advice." (Codes of Practice. Cattle, page 4).

This is largely because there is no clear understanding of stress and what it involves in the farm situation. A simple and practical solution is difficult since the measurement of stress on physiological grounds on the farm is not practical at present. Some studies using physiological measures of stress and correlating these with environmental changes which disturb behaviour have been carried out in recent years. Thus Kilgour & De Langen (1970) measured corticosteroid levels in sheep as the result of putting them in trucks, dipping, shearing and killing. They found that there was very considerable individual variation on corticosteroid levels, between treatment and between individual sheep. However, their results confused the effect of isolation and subjection to an unfamiliar environment, which in itself can be traumatic and induce changes (see page 38 et seq.), with the other experiences. Whittlestone et al. (1970) found an increase in cell count in bovine milk following isolation and the chasing of dairy cows. A similar increase was found following injection of ACTH, thus they assert that the former was a response to stress. Freeman (1971) measured ascorbic acid depletion in the blood serum of the domestic fowl under extremes of temperature, handling, shock, food and water deprivation and debeaking. Many more of these types of studies related to the everyday environment of the animal are needed. However, it must be realised that merely the taking of the blood sample itself can effect changes in blood steroids. Further work on the correlation of behavioural changes *with* physiological measures of stress could lead to behavioural measurements of stress at a subclinical level which would be more useful. The ambient causes are likely to be multi-factorial rather than related to a single factor as Fraser (1974) suggests.

In conclusion, it is fair to say that the use of the term 'stress' in relation to agricultural animals in their day to day environments is confusing rather than helpful.

CHAPTER 2
AGGRESSION

This term is generally used and misused in much the same way as 'stress'.
This is no place to review the innumerable ways in which it has been
used. Throughout this report, unless otherwise stated, 'aggressive
interactions' or encounters will be used to indicate the attack (by biting,
kicking and pecking) of one animal by another. This normally refers to
intra-specific encounters. An intention to attack which is indicated by
threatening postures or movements has also been used to indicate
'aggression'. It is reasonable to assume that threats represent an increase
in the tendency to attack and they will also be discussed here.

One of the most serious consequences of crowding is an increase in
'attack'. In rodents this increases under crowded and confined conditions
(Christian 1963 and Archer 1970a). It has also been shown in primates to
be related to increasing population density (e.g. Jay 1965) and
restriction (Mason 1960; Rowell 1967). Southwick (1966) reduced the
floor space for 17 rhesus from 1,000 to 500 sq.ft and found the number
of aggressive encounters greatly increased.

In intensively managed agricultural animals one of the main behavioural
defects that has been identified is an increase in aggressive interactions.
This is particularly apparent in pig and poultry units but is also of some
consequence in beef units (Hastie 1969) and with further intensification
it is likely to increase.

One of the most important restraints which increases attack in crowded
and confined environments is the lack of possibility of escape.

Lack of possibility of escape. When attacked, the response is usually to
move away. However, if this emigration is prevented, then the receiver of
the attack will either show typical fear responses, such as immobility, or
he will attack back. On the other hand, if the environment is provided
with areas to escape to then the attacked animal can flee or hide. For
example, Chance (1954) showed that the provision of a hide was enough
to reduce the occurrence of audogenic shock and hyper-excitement in
confined and crowded mice. Chance & Mackintosh (1962) showed that
just the provision of wood wool in the mice cage was enough to decrease
the number of aggressive encounters. Thus in pig units, for example,
hide-away areas in the pen could easily be constucted. Alternatively, a
double-storey pig house would allow pigs to get out of the aggressor's
vision, thereby possibly reducing aggressive encounters. It might also

reduce the price of buildings per pig.

Within the modern farming situation there are two common forms of attack on conspecifics. These have been the subject of some recent work and will be considered first. They are feather pecking in fowls and tail biting in pigs.

Feather pecking in fowls. Feather pecking in chickens, it has been argued, is not so much an 'aggressive response' as a response to dietary inadequacy or 'nutritional stress' (Wood-Gush 1971). However, feather pecking, cannibalism and tail biting in pigs involve an attack on other individuals and therefore will be included here, whatever their cause. Wood-Gush (personal comment) considers that feather pecking is not an attack since it is not directed to the head where attacks are directed. There is some evidence that inadequate diets, particularly in relation to protein levels, affect inter-pecking. For example, Schaible et al. (1947) found that pecking could be reduced in white leghorn chicks by the addition of casein to diets low in protein, crude fibre and phosphorus. Arginine reduced death from pecking in white leghorns (Miller & Bearse 1938). But Kuo (1960) found that high protein diets increased attacks in quail. Miller & Bearse's work also suggests that the *form* of the food is important in controlling cannibalism.

Hughes & Duncan (1972) in a recent experimental and review paper found that the strain of the bird, the housing system, the light intensity and position of the bird in the house had the largest effect on feather pecking, whereas the diet and the group size had minor effects. Very surprisingly, however, the population density had no effect. Their differences in space allowance varied only between 12.3 sq.m. and 14.2 sq.m. which, as they themselves point out, may be too small a difference. They quote other workers some of whom show a similar effect (e.g. Marsden 1966). Others, however, found that pecking increased in incidence and severity with crowding (e.g. Skolund & Palmer 1961). It is therefore likely that crowding increases the propensity to peck but the relative effects of other contributory factors on this remain unclear.

Tail biting in pigs. Less work has been done on tail and ear biting in pigs, and there is no review to date. Gadd (1967) did a survey of 300 pig farms and found that it appeared to be caused by a variety of factors, such as type of feeding, over-crowding, 'boredom', the presence or absence of bedding, mange and mites, temperature and ventilation. The latter was confirmed by Van Putten (1969) to cause outbreaks. Jericho & Church (1972) have so many variables between treatments in an investigation of cannibalism in pigs that the ambient conditions primarily responsible are not clear. All that is clear is that there are a host of environmental conditions that will effect an increase in this form of attack and one of them is overcrowding. There is some evidence of this from Ewbank & Meese (1971) and Bryant (1970 and 1971). These are some of the few studies which have actually measured behavioural responses. They found that there was an increase in the number of aggressive encounters as space per individual decreased and group size increased. Nevertheless, many more measurements of behavioural changes in different environmental conditions are needed before the relative effects of all these variables can be assessed.

A criticism of the studies on both feather pecking in fowls and tail biting in pigs is that no effort is made to measure other behaviours and their frequencies between treatments. For example, the measurement of an increase or decrease in activity, fear responses, vocalizations and other types of interactions between individuals would lead to greater under understanding.

The situations giving rise to attack and fear responses have been reviewed recently by Archer (1974). He concludes that it is a discrepancy between observed and expected phenomena which gives rise to both. This is no place to engage in theoretical arguments; for further details the reader is referred to Archer (loc.cit.). The types of situations which Archer points out can give rise to attack are considered briefly below, with particular reference to those frequently encountered by farm livestock.

Violation of an animal's *individual distance* is a common cause of aggression. This has been shown to be the case experimentally in birds (e.g. Andrew 1957 in buntings) and anecdotally reported in many mammals (e.g. Hediger 1955). It has not often been pointed out that individual distance is a dynamic concept, which will alter with the internal state of the animal and the ambient conditions. Thus the individual distance of a mare during oestrus with regard to the stallion is substantially reduced, although at such times her individual distance to her foal may *increase*. In crowded environments, the violation of an animal's individual distance results in attack.

Marler (1956) suggested that the concept of individual distance was the precursor of *territoriality*. Thus it became extended to include a whole area which was defended against intruders. This may continue through the whole year or, more commonly, be confined to the breeding season. Territoriality in the sense of a defended area has been shown in older bulls (Kilgour & Campion 1973) and Soay sheep (Grubb 1974).

The defended areas are not well defined in the younger bulls, however. Details on the establishment and maintenance of territories in domestic animals are lacking since domestic animal societies are rarely allowed to develop to such a point.

How the concept of individual distance can be extended to territoriality is easier to explain when it is considered that another situation which frequently gives rise to attack is the *introduction of unfamiliar animals* into a familiar environment. Fighting is uncommon within established groups of animals in a familiar area. However, the introduction of strange animals, particularly males, may induce fighting. This has been shown in the domestic fowl (Horridge 1970) and the rhesus (Southwick 1967), for example. Poole (1973) showed that attack in such situations was not confined to males since the introduction of females and young polecats into an established group also instigated attack. This has also been reported in pigs (Bryant 1971) and there is anecdotal evidence in cattle, horses, dogs and cats.

Attack by animals in *familiar environments towards unfamiliar objects* is also common (e.g. Baenninger 1973 in *Peromyscus,* Van Lawick-Goodall 1968 in chimps). An unfamiliar individual (or part thereof) of another species can also elicit attack (Horridge 1970 in chicks). Both attack and fear responses are common in farm animals when strange humans, for example, move among them as many veterinary surgeons and agricultural inspectors know to their cost. Since this attack wanes with time, it is suggested that it is the novelty of the object or person which elicits the attack (Archer loc. cit.).

Similarly, the placing of an animal, or group of familiar animals, into an *unfamiliar environment* may induce attack with the group (e.g. Willis 1966 in pigeons; Archer 1970b in mice; and Poole loc. cit. in polecats). Cattle, horses and pigs show similar responses when introduced to strange surroundings (Kiley unpublished data). However, other responses such as an increase in activity and fleeing or attempts to flee and defaecating may increase in such situations (Fraser 1974 in pigs and Kilgour in press).

Other situations normally evoking aggression which are well known and should be considered within the farm are *frustrated non-reward* (Berkowitz 1962, for further discussion) and *thwarting*, where the animals are physically prevented from obtaining a desired goal (Kiley 1969 in calves, horses, pigs, dogs and cats; and Duncan & Wood-Gush 1971 in chickens).

A particularly important type of thwarting which induces aggression within the farm situation is the *distribution and type of food.* If the distribution of the food is over a small area then, as Southwick (loc. cit.) showed in rhesus, aggressive encounters increased. In addition, anecdotal evidence suggests that restricting the feeding areas as in trough feeding, or feeding from a silage face, increases the numbers of aggressive encounters and interferes with the intake of individuals in dairy cattle, beef, poultry calves and goats (e.g. for goats, Syme et al. 1974).

However, the allowance per individual depends on the species and their social organization as well as their size and weight. Thus, with adequate supplies, cattle can be allowed between 22 cms to 30 cms trough space or eating face for each individual to obtain adequate rations. By contrast horses must be allowed several metres to ensure each individual eats at all (personal observation). Ewbank & Bryant (loc. cit.) found that one pig would stand in front of the trough after feeding preventing others obtaining the food. These were confined animals with rigid dominance hierarchies.

The type of food fed is also of importance in considering trough size. Highly palatable and sought after food increased aggressive interactions in a herd of goats, as did novelty of the food stuff (Syme et al. 1974).

Good intra-specific comparisons are not available to date but it is evident that homogeneity of sex, age and size of animals in a group is likely to contribute to each individual obtaining a ration, since no individual will have advantages in terms of size and weight. With such a group, a smaller feeding area per individual could be allowed. Homogeneous groups may not, however, reduce aggressive encounters *outside* the feeding situations (see page 28).

Low reinforcement schedules and such effects as *drug withdrawal,* for example, from morphine (Boska et al. 1966) may induce attack. All of these situations can be encountered on the farm. Archer (loc. cit) suggests that both aggressive and fear responses occur in similar situations. Kiley takes this further and suggests that there is an increase in many other activities as well in such situations (Kiley 1972 and in press).

Early social experience can also affect attack. The effect of isolation is, however, discussed elsewhere (page 38). Southwick (1959) showed that cross-fostered male mice from a non-aggressive strain to an aggressive mother showed substantial increases in aggression when compared with those fostered to their own strain. Levine (1957) showed that neonatal handling of rats produced an increased readiness to attack when tested as adults (see page 29).

Internal change and aggression. Internal changes can, of course, affect

attack. This, it is generally agreed, is mainly by changes in the levels
of excreted hormones. These will be briefly considered.

Testosterone and aggression. This hormone is usually considered
responsible for male-like behaviour and an increase in aggression in mice.
Male laboratory rats (Seward 1954), voles (Clark 1956) show greater
readiness to attack than females. Proximity-induced fighting and pain-
induced fighting is increased in rats by androgen treatment (Seward loc.
cit. and Ulrich 1965). Conversely, castration reduced readiness to attack
in proximity-induced fighting in red deer (Lincoln et al 1972) and mice
and rabbits. If androgen is injected into castrates, then there is an
increased readiness to attack. In farm animals one of the main reasons
for castration is to reduce aggression and thereby increase ease of
handling (e.g. stallions and bulls). However, caution must be observed
with this approach since the high levels of aggression shown by many
domestic male farm animals may be to some extent the result of the
conditions under which they are kept. (See page 60 et seq.).

Maternal aggression. Many females show an increase in attack towards
their own and other species before and after parturition which is
restricted to the period of infant protection (Mayer 1968). It has
been suggested that maternal aggression was associated with the
lactation period, implicating prolactin in its control. Bull-like behaviour
of pawing, head rubbing and threat posturing is common in beef single
suckler cattle just before and after parturition (Kiley unpublished data).
This suggests that changes in the levels of oestrogen and progesterone
may be equally important in this species. In hamsters, for example, high
levels of aggression have been related to progesterone (Payne &
Swanson 1972); oestradiol and ovarian implants can also increase
aggressiveness in this species (Vandenbergh (1971). Clearly there are
variations in the behavioural responses to different gonadotrophins
between species. To date farm animals have not been studied in this
context.

Pituitary-adrenocortical hormones. ACTH appears to be important in
controlling levels of aggression. Thus high levels increase attack while
low levels decrease it. However, these relationships, together with the
past experience of the animal, affect this in a complex way which is
not understood.

Pain. Ulrich (1966) and Azrin (1964) showed that the administration of
electric shocks to mice would make fighting more likely. Ulrich quotes
others who have shown pain-induced aggression in hamsters, mice, cats
and monkeys. This also occurs in dogs and horses (personal observation).

This, then, gives a brief summary of the types of situations giving rise to aggression. It is clear that much more work needs to be done on farm animals to assess the relevance of many of the conclusions derived from work with laboratory species. There is no doubt that, particularly with regard to the expression of aggressive responses, species differ greatly. However, it is important to be aware of the conclusions derived from such studies to act as a spring-board for the study of farm animal behaviour.

CHAPTER 3
CROWDING AND CONFINEMENT

A definition of 'intensive husbandry' could well be the husbandry of animals under crowded and confined conditions. Thus the behavioural effects of crowding and confinement are central to our theme.

Under unrestricted conditions, the normal response to, for example, violation of individual distance by another animal is avoidance and ultimately migration. However, under confinement there is a limit to this, and under crowded conditions this is no longer possible. The results are, of course, an increased contact between individuals. This increased contact between individuals is not necessarily detrimental to production or health, it may merely result in changes in the social organization and the development and maintenance of bonds between individuals. For example, in a captive group of eland there was found to be a large increase in the amount of amicable contact (licking and rubbing) when compared with the wild group (Kiley 1974). There are, however, few studies to date which have looked at all types of interactions between individuals in one environment, let alone as a comparison between two. The detrimental effects of such increased contact between individuals are better known and one of the major effects is an increase in attack.

Crowding. The physiological effects of increasing population density have been most carefully studied in rodents (Crew & Mirskaia 1931; Southwick & Bland 1959; Christian 1963; Christian & Davis 1964; Antrum & Von Holst 1968; and reviewed by Myers 1966). Reviews of the behavioural effects of increased population have been written by Christian 1963; Christian and Davis 1964; Thiessen 1964; Myers 1966; Archer 1970a. Although there is little work to date in detail on other mammals, similar types of physiological and behavioural responses have been shown to occur in, for example, the sika deer (Christian 1963), primates (Rowell 1967; Jay 1965 and Sassenrath 1970) and chickens (Siegel 1959).

From these studies the main behavioural changes that emerge are listed below, with the page references for their discussion:
(1) An increase in aggression or attack (page 8 et seq.).
(2) Changes in the organization of dominance, with a tendency towards the establishment of a more rigid dominance hierarchy, or total social disorganisation (page 18 et seq.).
(3) Reproductive defects including infertility. Whether this is primarily

because of social factors changing physiology or direct physiological effects of crowding is obscure (page 12). In addition, on the farm crowding can result in poorer management which can cause a drop in fertility (e.g. non-recognition of oestrus, page 47 et seq.).
(4) Deficient maternal behaviour that increases infantile mortality (e.g. non-recognition of young, page 53 et seq.).
(5) Changes in general activity levels, with their concomitant problems (page 28).

All these behavioural changes can also occur with other environmental factors, with which they have often been confounded: for example, the population size and the size of group and its composition, the size of the feeding trough, the type of feed, the temperature, ventilation and shape of the pen, the complexity of the environment and so on. There are few studies where the effect of space per se has not been confounded by several of these other factors (e.g. Ewbank & Bryant 1969).

Further, the space requirements per individual are bound to vary as the other environmental factors vary. Thus it is not possible to give minimum space requirements for the animal in terms of square metres per kilogram live weight, although guide lines are useful (e.g. Codes of Practice Ministry of Agriculture, Fisheries and Food 1972). The inter-relationships of these various environmental factors have been pointed out previously. Christian (1964) states that the density was not only related to the numbers of animals per unit area but also to the 'socio-physiological pressures' (this is presumably what others mean by social stress, see page 4 et seq.), and that this was not a static state, but rather a constantly changing one in response to the environmental changes that occur.

An experimental demonstration of this is given by Southwick (1966) who found that when the size of the living area for his rhesus monkeys was reduced by a half (1,000 to 500 sq. ft) aggression was increased. However, this increase in aggression was less than that seen when strange monkeys were introduced to the group. Here, therefore, social rather than spatial considerations may be more important in the exhibition of aggression.

Another factor that must be considered when thinking of overcrowding is that, although many agricultural animals have been domesticated for several centuries, they are likely to adapt most easily to the type of environment which morphologically and behaviourally they are best suited to, that is, that for which they evolved to live in. Thus social species should be kept as groups, and others should be solitary. Again, this is not a new idea. Eisenberg (1967) has postulated that different species have different thresholds of responsiveness to grouping depending on whether they are solitary, semi-solitary or social in the wild.

Confinement. The distinction between crowding and confinement and the behavioural changes related to these is not clear. No experimental work has been reported to date.

The farmers' approach to such problems is pragmatic; harmful effects of crowding or confinement are usually treated by further restriction. Thus sows are put into farrowing crates or stalls, laying hens in battery cages, calves in crates and mothers isolated from young. Of course, there are other reasons why the animals are restricted or isolated in these ways, for example in order to be able to keep individual production records (e.g. laying hens) and ration feeding or prevention of intersuckling in calves. However, it must not be overlooked that such isolation or further restriction may itself enhance the problem or create new ones.

Thus the effects of *isolation* and *social facilitation* of behaviour must be considered (page 38).

A further effect of such restricted animals is the development of stereotypies, some of which can become pathological and many may cause drops in production. *Stereotypies* involve repeated actions such as crib biting, fleece chewing, self licking, weaving and intersuckling. A review of their causes and effects is therefore included (page 74).

In the future it is likely that further intensification of animal husbandry will develop. It is therefore necessary to examine the causes and effects of such changes in order to be able to help with present and future housing design and production.

CHAPTER 4
AN EXAMINATION OF 'DOMINANCE'

There are several reasons why we should discuss dominance. Firstly, this term is used widely, both in the scientific and lay literature as an explanatory term, whereas it has only descriptive value. Secondly, it is important at this stage to examine what is meant by it and whether it is, in fact, a useful term to use. A clearer understanding of what is meant by dominance and how important it is in animal societies will help husbandry and building design considerably.

It is usually considered that 'priority of access to any finite resource' defines dominance, and that dominance gives priority of access, a circular argument from which, as Gartlan (1968) points out in his review of the subject in primates, "all that can be properly inferred (from this) is that dominance is priority of access". Frequently aggression is used as synonymous with dominance.

A recent definition which combines these two definitions of dominance is Bouissou & Signoret's (1970) of a dominant animal which is " celui qui peut attaquer impunement les autres, et jouit de priorite lorsqu'il y a competition". Others use dominance when referring to the larger animals (e.g. Stephens 1974). Yet others do not define what they mean or have measured when they use the term (e.g. Short 1970; Joubert 1974; Grubb 1974).

The concept of dominance has been used since Zuckerman (1932) as one of the most important factors in social structure. The idea became even more acceptable when it was suggested that it reduced aggression (Scott 1956). This was particularly important since ethology has, until recently, been dominated by theorists who believed in innate aggressive drives.

The concept was formulated as a result of work with captive animals and in particular primates. Gartlan (1968) showed, however, in *Cercompithicus aethiops* that aggression is lower in the wild and the hierarchy is very much less rigid or obvious than in the same species in captivity. Preliminary studies on beef cattle at pasture compared with housed animals show similar effects (Nicholson 1974). In effect, then, the rigid establishment of a dominance hierarchy (in so far as it exists) may well be an artefact of confinement or captivity, where aggression is often increased (page 8 et seq.).

The original work that gave rise to the idea of peck order and dominance hierarchy was done on a small group of chickens (Schjelderup-

Ebbe 1931) in which straight line hierarchies were described. It became
increasingly apparent with further studies that a straight line hierarchy
was insufficient to describe the complex relationships between individuals
where A dominates B, who dominates C, but C dominates A. Instead of
a reappraisal of the concept of dominance as a result of such findings,
the idea was merely expanded to accommodate such difficulties to a
degree where very elaborate formulae were used to allow assessment of
the 'dominance value' of an individual. The zenith of this elaboration is
shown in Beilharz & Mylrea (1963) where the dominance value of the
animal is represented as an angle. The alternative approach was to dismiss
the idea that dominance hierarchies could be circular and make complete
assumptions concerning the position of an individual in a linear hierarchy,
on the basis of as few as two encounters of that individual with *any*
other (Schein & Fohrman 1964).

Baenninger (1966) and Bernstein (1970), in rodents and primates
respectively, have reviewed the literature and made a critical assessment
of the concept of dominance. An attempt will now be made to extend
this approach to the considerable body of work on domestic animals, in
particular dairy cattle.

A. The measurement of dominance
With a concept as widely used in psychological, agricultural, behavioural
and wildlife literature as dominance, it is frequently assumed that
standard rigid methods of measuring it have been developed. This is, in
fact, not the case.

In the first instance different people have measured different behaviours
from which to assess dominance hierarchy. For example, some have
measured threats (e.g. Collias 1950, Guhl & Atkeson 1959); others
encounters won (e.g. Beilharz & Mylrea 1963); yet others have measured
amount of time feeding at a restricted food resource (e.g. Candland &
Bloomquist 1965, Beilharz & Cox 1967); others withdrawal (e.g. Bryant
1971); others avoidance (e.g. McBride et al. 1964, among other measures)
and others grooming (Spigel & Fraser 1974, but see Drews 1973). Most
commonly a glorious hodge-podge of all or many of these variables has
been recorded and summated for an assessment of dominance (e.g.
Brantas 1968a, Rowell 1966, Schein & Fohrman 1954).

Whether or not the measurement of aggressive encounters or withdrawal
gives similar dominance hierarchies has rarely been tested. Where it has,
different results are given by different species (e.g. grouse, Marquiss 1974;
and eland, Kiley 1974).

Secondly, there are two major situations or methods of collecting the data. These are:-
i) By direct observation of the animals in the field and the outcome of naturally occurring contests between individuals. Needless to say, this is a very lengthy and time consuming process. In addition, these contests may not be directly comparable since they may well be about different things, for example, right of way or access to food or drink, or access to female or male, young or litter mate. Another disadvantage of this method of assessing dominance is that encounters between every possible combination of individual, particularly in a natural population or large group, are unlikely to take place. Indeed, if quantitative measures based on encounters were to be kept, the number of observations which would have to be made to justify any valid conclusions is very high even in a small group (see below). Nevertheless, this is a method that has been widely used in assessing dominance (e.g. Schein & Fohrman 1954; Grant & Chance 1958; Kilgour & Scott 1959; McPhee et al. 1964; Rowell 1966; and Baenninger 1966).

Frequently, in observational studies of wild animals in particular, the dominance hierarchy within the particular population studies is merely reported. No results or methodology are given (e.g. Short 1970; Grubb 1974; Joubert 1974).

It is precisely under these circumstances (or with domestic animals, those at pasture) where one would expect a dominance hierarchy to be less obvious or perhaps not to exist at all (e.g. Ewbank 1973). In the grazing herbivore, where natural resources (e.g. food and water) are scattered and usually equally available to all individuals, there is little reason to suppose that conflict between animals is of prime importance in their social organization.

ii) By staging encounters between pairs of individuals isolated from the group and often in strange surroundings, the numbers of wins and losses can be established as a result of contests for food or water. This method clearly has the advantage that each pair can have a staged encounter, and the amount of data collected for the time expended is greatly increased (Collias 1950; Candland & Bloomquist 1965). However, under these conditions it is not at all clear what is being measured other than the ability to gain access to food or water in an isolated pair encounter. What relevance, if any, this has in situations more normally encountered by the animal is questionable. Also the experience of having lost or won a previous encounter in such a situation affects the results (Ginsburg & Allee 1942).

It is likely that the measurement of dominance by either of these methods is bound to show differences, since the animals in the second situation are isolated and in a strange environment. No comprehensive studies using both methods with a similar population have been made, although Bouissou (1970) attempted this in cattle. She gives no details of how the hierarchical relationships were arrived at observationally (method i) to compare with those in an experimental situation (method ii). Grant & Chance (loc. cit.) used method (i) and found that dominance orders in rats remained the same from observation to observation, whereas Candland & Bloomquist (1965) used method (ii) and found that they changed from observation to observation. Nevertheless, this has not prevented the use of a combination of the two methods to arrive at dominance hierarchies (e.g. Kilgour & Scott 1959).

Both methods require a very large number of observations to be made since at least 10 encounters between every possible paired combinations of animals in the same type of situation is necessary for valid statistical treatment. This is only possible when the animals are divided into small groups of from 5 to 10 individuals which cuts down the number of observations required. As pointed out by Gartlan and demonstrated in the chick by Guhl (1953), small groups tend to have a more rigid form of hierarchy, which may bear no relation to the type of organization of the hierarchy in the larger group in which the animals usually associate or are kept. Schein & Fohrman (1965) tried to work out the hierarchical relationships of 163 dairy cattle and heifers. They made over 5,000 observations but, as they themselves point out, in order to have only one observation on each possible combination of each animal with each other, they would have had to have 13,000 observations. Since at least, say, 10 observations per encounter are necessary for valid statistical treatment, this makes the total number of observations needed 130,000! Not only did they have as few as 2 observations on some pairs, but they also collected this data from animals that were moving from herd to herd. The total herd structure was fluctuating continually and contact with quite strange animals could occur at different times.

Thus the methods of measuring dominance that have been used are open to severe criticism. Since the results of the studies that have been done can be questioned on these grounds, it is hardly necessary to criticise the concept on others. Nevertheless, so many people have studied dominance hierarchies in small groups of animals that the uses to which the concept has been put deserves further consideration.

B. Relations of dominance to an individual's physical characteristics and production

Several workers have related dominance values in mammals to various physical characteristics such as sex (Zuckerman 1932), age (Guhl & Allee 1944), weight (Dickson et al. 1967), height (McPhee et al.1964), girth (Beilharz & Mylrea 1963), presence or absence of horns (Woodbury 1941), breed differences (Wagnon et al. 1966; Guhl & Atkeson 1959). In hens, breed differences (Tindell & Craig 1959), body size (Ewbank 1969), length of time in the group (Schein & Fohrman 1954) and egg production (McBride 1960) have all been studied. Each of these workers finds that the particular characteristic selected by them correlated with the rank order better than those selected by other workers!

However, there does seem to be some agreement that milk production does not correlate with dominance order in dairy cattle (Schein & Fohrman 1954; Dietrich et al. 1965). Guhl (1953) and Sanctuary (1932) found that dominance was related to egg production in small groups of chickens, whereas James & Foenander (1961) estimated social rank from paired encounters of laying hens and found that there was no effect of rank on production, but time of commencement of egg laying was correlated with rank. McBride et al. (1964) found that the heavier pigs at weaning were of higher social rank (they weighed 12 pounds more at 12 weeks old than the subordinates), and suggested that weight gain may be an indicator of social status. Caution should be used with such an approach, to avoid circular arguments. Thus Stephens (1974) defines dominance in calves in terms of weight (the heavier, the more dominant).

Beilharz & Cox (1967) found that dominance 'value' correlated with growth rate and size in pigs, whereas Rasmussen et al. (1962) found no meaningful relationship between rank order and body weight in gilts.

Dominance status in confined environments or with restricted feeding face can affect intake, and thus production. Ewbank & Bryant (1969) found that some pigs had priority of access to the feed and when satiated did not move away, thus preventing others having any access. Larsen(1963) found that subordinate cattle spent more time actively moving about, 106 to 179 minutes more per day, and 25 to 55 minutes longer eating than the dominant animals. Here again, however, details of measures of dominance are not given. It appears that those spending more time eating were defined as the subordinate.

C. Recognition of dominance by the animal

It has often been assumed that the recognition of an animal's dominance status by another is the result of individual recognition of that animal by visual cues. However, few experimental studies of this (other than on poultry) appear to have been made until the recent work of Bouissou 1970).

She found that when two animals were tested in a food-competitive situation the subordinate fed more if contact with the head was prevented between the two cows. Deprivation of sight, tested in the same way, had little effect on social order as previously assessed. This result can be interpreted to suggest that the immediate recognition by one animal of another's position in the hierarchy does not exist unless reinforced by head contact (i.e. butting).

Meese (1971) attempted to isolate the ways individual pigs recognised each other and thus their dominance status. He also found that sight was not important but that hoods over the face, which he suggests prevented identification by pheromones secreted from glands on the face, resulted in non-recognition of individuals by others of the group.

D. Relationship of dominance order to internal factors
Dominance has also been related to stage of lactation and gestation in the dairy cow and primates (e.g. Dickson et al. 1967). Beilharz & Mylrea (1963) maintain that the pregnant animals are more submissive. There is disagreement here, since Donaldson & James (1964), in dairy cows, show that pregnant animals tend to become more dominant. Aggressive interactions increase in beef cattle around parturition (Kiley, in preparation), which could be interpreted as supporting the latter view. Others have remarked on the likelihood of the hormonal and temperamental status affecting the dominance hierarchy (Woodbury 1941).

There is no doubt that some hormones (e.g. testosterone) affect the degree of aggressiveness displayed by an individual and thus possibly affect the dominance hierarchy. Short (1970), for example, working with wild red deer, found that testosterone-implanted males rose to a higher position in the dominance hierarchy, even when implanted out of the rut, but again he gives no details on how the dominance was measured.

E. Other physiological effects of dominance
In captive or crowded societies of animals it has been shown that there are various physiological changes in the animal which are similar to those described by Selye (1950) as the result of stress. It has also been shown in rodents that these stress effects are particularly shown by the subordinate animals. They can be recognised by enlargement of the

adrenals (Christian 1963) or depletion in ascorbic acid levels (Archer 1970c), and occur in the types of societies in which the rigid dominance hierarchies have been described (Schjelderup-Ebbe 1931; Zuckerman 1932; Rowell 1966). This has lead various people (e.g. Gartlan 1968) to suggest that this type of dominance hierarchy is a social artefact, symptomatic of social stress.

Many agricultural animals are now kept under conditions of crowding and confinement (pigs, beef cattle and poultry and they are likely to be introduced in sheep, Owen 1969). It is not surprising to find these animals forming rigid dominance hierarchies associated with increases in aggression. From the point of view of production (i.e. growth rate and economics) it is likely that the more subordinate animals will suffer most under such a system. James (1949) showed that subordinate puppies grew more slowly than dominant ones. He suggested that the emotional disturbance and conflict in the submissive animals produce detrimental effects on growth. He does not, however, appear to have controlled for the reduced food intake in subordinate animals. In crowded pigs reduced food intake in the subordinates has been shown by Ewbank & Bryant (1969). They reported that the dominant animals stood in front of the trough and prevented the subordinate animals from feeding. A similar behaviour is a problem in dairy cattle self feeding silage or trough fed with inadequate trough length. In addition, Fraser (1961) reports that there were more internal parasites in the subordinate goats. The reduced resistance to disease and infection is another classic effect of stress (Selye 1950 and page 4 et seq.).

In the grazing animal, however, where it is possible to move away from the dominant animal, if it is recognised as such, such effects are not likely to be so obvious, and the hierarchy is less rigid (e.g. Wagnon 1965 with range cattle).

F. The relevance of dominance hierarchies to situations other than those in which they were assessed
It has been assumed too often that, if a dominance hierarchy was measured in one situation, it would remain constant in all situations in which there was competition for a resource. For example, it has been assumed that a male which obtained access to food in competition with others would necessarily copulate more and beget more offspring than others. Jay (1965), with langurs, showed that the copulation success of males was not correlated with their rank order, measured by threatening, presenting, grooming and soliciting females.

Jolly (1967), studying lemurs, found the same. A closer look at wild
societies of primates (Gartlan 1968) and captive baboons (Rowell 1966)
suggests that to generalise concerning dominance status from one
situation to another is unwarranted. Less detailed work of this type has
been done on ungulates. However, Wagnon (1965) observed cattle on the
range and found that the dominance order was constantly changing and
was not correlated with weight or age; nor was there a straight line
hierarchy. Ross & Scott (1949) found that even in a small herd of
confined goats dominance varied from situation to situation.

Furthermore, in dairy cattle there have been attempts to relate
dominance (measured by observation in aggressive encounters) to
leadership (Kilgour & Scott 1959), order in entering the crush (Donaldson
& James 1964) or order of entering the milking byre (Dietrich et al.
1965). No correlation was found, although in the case of milking order a
constant order was found; this is also true of sheep (Scott 1945).
Similarly, it has been observed in feral horses (Kiley 1965) that although
the stallion herds the mares when threatened by an intruding male, he
is dominated by the females when there is a conflict over right of way,
food, drinking resource or approach to young.

Thus it appears that the dominance hierarchy, in so far as it exists at
all in either more natural free range populations or even in captive or
confined animals, changes from situation to situation. In effect, it
appears that it might be more useful to describe the society in terms of
the roles of individuals, as suggested by Gartlan (1968) and Bernstein
(1970) for primate societies, rather than in terms of a nebulous
dominance. For example, individuals having different roles related to
leadership, crush order, milking order, attendant to oestral animal, access
to water, food or calves, butting order and so on.

Early in the history of behaviour, naturalists made attempts to assess
the personality of an individual cow, often scoring them from sanguine to
tranquil in temperament. It is suggested that a sophistication of this
approach in which the role or status of the animal is assessed in each of
the above types of situations (and others) might be more helpful in
relating the social organisation to production, rather than the somewhat
muddled assumption of dominance hierarchies. Such an approach has
recently been attempted by Kiley (1974) who assessed the rank of a
group of captive eland separately for each individual for threats,
withdrawing, affiliative behaviour, displays, general interest, social
involvement and other indices.

It has been pointed out that different environments can affect social structure. Gartlan (loc. cit.) goes so far as to say that social structure varies according to habitat rather than species in primates. Evidence for this can be found in Jay's work (1965) with different populations of *Macaca mulatta,* and Gartlan (1968) in *Cercompithicus aethiops.* Buechner (1961), with a high density population of *Kobus adenota* in Uganda, found an elaborate territorial structure and dominance hierarchy. Populations of kob in the Park Nationale Albert, where the density is lower, do not have the same territorial structure. So ungulate social structure can apparently change as the result of differences in population density and, probably, habitat.

G. Effects due to subordinate animal
As Rowell (1966) and Gartlan (1968) point out, it is not aggression on the part of the dominant animal so much as withdrawal by the subordinate animal that allows the identification of the dominance order. Rowell goes so far as to say that it is primarily the behaviour of the subordinate that controls the establishment of rank. This work confirms Seward (1945) who showed that the subordinate rat after losing an encounter, was less likely to fight at all, whatever rat he encountered.

H. Past experience, learning and dominance
Donaldson (1967) showed that past experience has an effect on dominance status in calves. Group-reared calves that were competitively fed were less dominant than those fed independently, that is, they had learnt to avoid others, whereas the latter had not. Calves raised and fed in groups were the most submissive animals, whereas those reared and fed in isolation were both less dominant and less submissive (that is, less socially involved). McBride (1964) found that the effect of one peck on a chicken was to prevent the chicken approaching another for one month. Work with rats indicates that past experience (i.e. winning or losing encounters) has a profound effect on the performance thereafter. If the rat won previously he is more likely to win again (that is, more likely to be dominant) and vice versa (e.g. Ginsburg & Allee 1942). These authors found that it was possible to condition the dominant animal to behave in a subordinate way but more difficult the other way around. Work with cattle and horses suggests a similar conclusion (Kiley, personal experience). Drugs can help with this, thus Beeman & Allee (1945) showed that thiamine increased social rank in mice and Bainbridge (1969) found that chloropromazine would increase the number of wins of a rat in social encounters.

An understanding of dominance and what it involves in the various agricultural species is very important to allow for each animal to receive its food ration at a restricted food resource, for example, and to enable better building design and facilities. However, as it stands at the moment, there is little agreement in what dominance involves, how important it is and how to measure it. There has been too much generalisation between species and situations. For example, there is no reason to believe that dominance must exist in the same form between two species of ungulates, where one is more social than the other (e.g. pigs (less social) and cattle). Even in two social ungulates, there are suggestions that the operation defining priority of access is different. For example, when cattle are fed at a restricted food resource, if they are provided with at least 9 inches of feed space, all will normally obtain a ration. Such a feeding space is totally inadequate for a group of horses or ponies which, to ensure some animals obtaining any food at all, must be fed at distances of several yards (personal observation).

A greater understanding of the animals and their *total* social organisation (which must encompass measures of personality) could be followed by the profitable application of such discoveries to animal husbandry. This will come from measures such as the total involvement of individual animals within the group, affiliative relationships and performance of other activities, rather than the simplistic observation of animals in conflict, which often has to be induced by the experimenter.

CHAPTER 5
ACTIVITY AND THE EFFECTS OF ENVIRONMENTAL CHANGE
AND COMPLEXITY

A. Activity
There are conflicting results on the effect of crowding on activity. Lloyd
& Christian (1967) provide evidence to indicate that in mice, as crowding
increases, so the activity of most individuals decreases. This would agree
with Morris' (1963) suggestion that zoo animals are more 'lethargic'.
Housed pigs certainly show less activity than ones on pasture (Ewbank
1973 and personal observation. Archer (1970a) suggests that changes in
activity are related to social status. He mentions several works on rodents
that indicate that, with crowding, activity of dominant animals increases
while that of subordinates may be greatly reduced (Clarke 1956 in voles;
Calhoun 1963 in rats).

There is little work on changes in activity in farm animals with
increasing crowding, although Heitman et al. (1961) found that in pigs
as the population density increased so more time was spent standing and
walking and less sleeping and resting. Similarly little is known of the
effect of population size (i.e. group size) on activity.

Once we have some measure of the normal amount of activity in farm
animals on the range, it is possible that gross changes in the level of
activity in housed animals could become one behavioural measure of
physiological stress (see section on stress).

B. Environment complexity
A complex environment can be considered in terms of either a complex
social environment, involving opportunity for social relations with other
conspecifics, or in terms of complex *physical* environment. The animal on
free range normally has a fairly complex environment in both the
physical and social sense. At the other extreme is isolation from
conspecifics in a dull, uniform and confined area. The profound effect of
this type of environment on behaviour has been demonstrated in
particular in primates (Chance & Mackintosh 1962; Draper 1965;
Sackett 1965 and is discussed below, page 38). Here we will discuss the
effects of a monotonous, 'dull' environment.

There is some evidence to suggest that an increase in the environmental
complexity, even just by the addition of wood wool in a mouse cage
Chance & Mackintosh 1962), reduces the number of aggressive

encounters. Similarly, giving mice the opportunity to climb and get away from the aggressor reduces the number of aggressive encounters. Chance (1954) showed that audiogenic hyper-excitement in rats can be suppressed by the provision of a hide.

The environmental complexity can be increased by handling animals. This has received much attention in rats and mice. Levine (1957) and Ader & Conklin (1963) found that handling of infants ensured that they showed fewer emotional disturbances when adult, although this effect may have been the result of changes in behaviour of the mothers after infant handling. Denenberg & Whimbey (1963) showed that variations in husbandry do have a very significant effect on later behaviour and physiology. They kept one group of rats on the same shavings and without handling, whilst the other group was handled and shavings were regularly changed. As a result, the handled rats weighed more and were more active. Whether these effects were due to early handling, or to changes in maternal behaviour following handling, is not clear, but either way there was an effect.

Krech et al. (1960) show that with rats in environments differing in complexity and training, the cortical to subcortical acetyl-cholinesterase ratio in the brain is lower in rats kept in the dull environment. Thus not only can a more complex environment affect behaviour in the rat (such as ability to learn and E.E.G. activity) but it also affects brain structure and organisation (Bennett et al. 1964).

In the husbandry of agriculture animals there is a constant trend towards reducing environmental complexity; this often correlates with labour-saving environments and a decrease in capital expense. Thus sows are now often kept in stalls facing a white wall. Such animals are not only very rigidly confined, but cannot experience any substantial change in visual stimuli. All other potential sources of stimulus change are also removed (locomotion, social contact, performance of different actions and so on). It is likely that conditions as extreme as this may effect some changes in brain structure. Direct effects on meat production are unlikely but not impossible. It is possible that these animals may eat less or distribute meals less effectively than free-range or animals in barns. It is probable that such treatment will affect reproductive performance and maternal behaviour (see relevant sections). In dull, monotonous environments, even with social companions, stereotypies also become more common (Meyer-Holzapfel 1968, see page 75).

Increasing the environmental complexity in the social sense is not often practical on farms. Introducing more animals of different age groups or sexes may have to be avoided for various reasons, and also is not always wise where conditions of group size and space requirements (see relevant sections) must be considered. However, increases in complexity of the physical environment would be possible if these were shown to be desirable, for example, providing hide areas and partitions, different floor levels for pigs, perches for laying hens. McBride (personal comment), for example, found that curtain partitions in laying hens' houses reduced hysteria (rushing about and smothering).

C. Environmental change

An environmental change can be either in the physical or social environment. Let us first consider:-

(i) Changes in the social environment. This involves the introduction of new individuals and the mixing of batches of individuals unknown to each other, or removal of familiar animals from the group.

In caged mice, rats and primates, it has been shown that the introduction of new individuals disrupts the established social 'hierarchy' (cf. page 18) and results in an increase of aggression and social disorganisation. Pigs are frequently sorted into different groups from different litters despite the opinion generally held by most farmers that mixing groups reduces weight gain and growth rate. Teague & Griefe (1961) showed that re-sorting of pigs every 28 days resulted in a considerable drop in weight gain. It has been shown that mixing groups of mice in new cages gave rise to wet faeces and other responses used as measures of fear. The mice which were in their own cages did not show such responses. Such findings are relevant to the difficulty often experienced by farmers who buy in weaner piglets from various farms. Loss of weight and disease is often encountered with such husbandry. Causally both these effects are related to the traumatic experience of being separated from the mother, moved from farm to farm or market and eventually placed in a strange environment, often with strange pigs.

In beef or dairy cattle, horses and zoo animals, the introduction of unknown individuals into groups causes social disorganisation and general disruption of the group. Schein et al. (1955), for example, showed a drop in milk yield of 5% on the introduction of a new member to the herd. This may well be proportional to the degree to which the animals are confined. It causes far less of a problem in animals in a grazing situation (personal observation and experience of many farmers and zoo keepers).

Such social disruption can cause injury particularly to the introduced

animal. However, as might be expected from an understanding of the
concepts of home range and territory, injuries and apparent social
disruption can be reduced by introducing the strange animals to each
other on mutually unknown ground (personal observations on cattle,
horses, pigs and dogs). Alternatively, introductions should be performed
in large areas wherever possible in order to prevent enforced encounters
between unfamiliar animals until social relations are established.

(ii) Changes in the physical environment. Placing guinea pigs and other
mammals in novel invironments, such as new cages, gives rise to fear
responses (Pearson 1970). Loud noises and sudden increased illumination
lead to stampeding and smothering in intensively housed poultry. These
reactions can become pathological. Thus animals in a house may show
prolonged immobility to only very slight environmental changes (e.g.
the movement of a person or noise of a bucket, Ferguson 1968). Such
behaviour may be detrimental to production. Whittlestone et al. (1970)
showed that environmental change (thunderstorms, chasing by dogs and
isolation) reduces milk let down. Changes in familiar routines can cause
considerable drop in production. Thus Kilgour (1969) found that
changing the position and side of milking reduced the milk yield by 20%.

Immobility responses are usual in housed pigs. The entrance of the
pigman and, in particular, of a stranger is enough to trigger this response
throughout the house. Alternatively, hyper-excitement may be triggered
off in dull environments by similar environmental changes, resulting in
stampedes or sudden leaping about (e.g. in pigs and veal calves).
Hysteria in fowl (mass flying and running) is a particularly harmful
example of such behaviour.

Because these over-reactions often cause injury, it has been argued that
the type of environment that should be aimed for is one in which all
extraneous stimuli and environmental change is reduced (for example, by
the use of sound-proofed houses), where feeding is automatic and there is
a minimum of any type of disturbance (Fox 1968). This approach would
appear to be particularly expensive in terms of capital investment and,
more important, is quite unlikely to solve the problem. One of the major
problems for confined and crowded animals in monotonous environments
is the absence of novel stimuli. It appears that, as a result, the threshold
for reaction to any available stimulus drops. Since it is impossible to
eliminate all stimuli from the environment, to reduce the number and
amplitude of those present may well result in a further drop in the
threshold for reaction, and thus increase the problem. The answer may
well be to *increase* the variety of auditory and visual stimuli rather than

to decrease it further.

D. Adaptation to environmental change

That an environmental change is sought after is illustrated by work with fowl (Duncan & Hughes 1972) where it has been shown that these animals will work to produce an environmental change.

Past experience clearly affects the likelihood of such over-reactions due to sudden environmental change. Thus horses, sheep and cattle can rapidly adapt to frequently changing social and physical environments. This is seen in livestock taken to shows or competitions. These animals can be gradually conditioned to be transported, subjected to loud noises, bright lights and constantly changing social companions and yet perform adequately and quietly. On the other hand, inexperienced animals such as moorland sheep reared by mountain-mothers show symptoms of environmental stress (increase in plasma corticosteroid levels) even when penned and handled, and some of them may die (Dr Foot, H.F.R.O. personal comment). It has been said that intensive pig units do well adjacent to airports in constant use, where the noise level is constantly fairly high. Where there is sudden unexpected high noise levels, such as near a seldom used airport, pig units are unsuccessful.

These examples suggest that animals can adapt to considerably different sensory stimulus levels as long as they are relatively constant. If the stimulus change is gross, sudden and unexpected then the behavioural responses may well be detrimental to production.

E. Control of levels of environmental stimulation

It appears therefore that either too much or too little environmental stimulus is likely to have deleterious effects on both physiology and behaviour of the animal. This leads to the theory of an optimum level of environmental stimulus, above or below which various physiological changes (usually related to 'stress', e.g. corticosteroid secretion) take place (e.g. Welsh 1964). Thus, too much environmental stimulation in crowded and confined environments and too little in monotonous, dull, simple and isolated environments is likely to result in physiological responses of stress, drop in production and an increase in disease (see section on stress).

Within the modern farm environment both extremes of environmental stimulus are prevalent. Firstly, to overcome the effects of crowding and related social factors, better use of the available space should be offered, as well as control of such factors as size of group, and the introduction of strange animals.

Secondly, there are three major ways in which levels of environmental stimulation could be increased.

(i) Direct increase in slightly changing environmental stimuli

(a) Visual stimuli. The increase in visual stimuli could be achieved by the provision of double glazed windows in, for example, chicken and pig houses, which are normally without windows out of which the animals can see. It will be argued that this will increase the cost of building construction. This cost should be set against the possible reduction in production in monotonous environments. Such a suggestion obviously needs experimentation.

There are some classes of livestock where it would be feasible to provide outdoor uncovered yards where they could idle (e.g. dairy cattle, which are zero grazed, sows and beef animals). Dairy cattle apparently may choose to go and stand in the rain if they are normally kept under cover (anecdotal evidence from farmers). The heat loss from the buildings could be controlled by having a swing door that the animals operate themselves. Another possibility is the provision of a mirror to allow animals to watch themselves or others. Pigs will very rapidly learn to look up in a mirror placed above them (personal experience). In the laboratory isolated chicks are often given a mirror which appears to facilitate feeding (Prof. R.J. Andrew, personal comment). A mirror might be therapeutic to some animals kept in isolation (e.g. bulls). Again, we have no information as to the possible effects on production.

(b) Auditory stimuli. A very simple way of increasing slightly changing auditory stimuli would be the provision of tape recorded music through the buildings. It is reported by farmers who use this method that music tends to increase milk let down and relax the cows in the milking parlour. Its wider use might prove useful and cheap. At present intensive husbandry units are often presented with white noise from fans or other automatic machines. It is likely that such a background noise will increase the reaction to extraneous noise, when it is heard, and thus help promote smothering in hens and other panic reactions. The provision of windows in the building would also allow the noises of the farm yard to penetrate the enclosed animal house. Sudden loud noises should, however, be avoided, although animals will habituate to these if continued for long enough (Kiley, unpublished data from a veal unit).

(ii) Occupational therapy. The provision of objects to manipulate and other things to do appears to have beneficial effects in institutionalised human beings and zoo animals (Hediger 1955). In the farm situation

little effort in this direction has been made to date. Sometimes tyres or balls are provided when pigs indulge in tail biting. This is not always successful in reducing tail biting (Mr Frith's farm, Fletching, Sussex) but the age at which the plaything becomes available may be crucial. There are a variety of such objects which could be provided at very little expense and which might well help towards increasing the level of environmental stimulation.

(iii) Operant conditioning and manipulation of their own environment. One of the hazards of animals housed in controlled environments is that they are prevented from performing many of the normal activities which would occupy them during the day outside. Thus, their food is provided, they do not have to search for it or even graze it, water is also provided. They do not have to take behavioural measures to keep themselves at a reasonable temperature. It is suggested that the provision of methods for the animals to perform some of these tasks for themselves would help to keep the environmental stimulation level at its optimum. In addition, if suitably designed, such measures could cut down labour costs.

We know surprisingly little about the ability of farm animals to learn 'operant responses' (see page 71). However, it has been shown that chicks will work for food (Duncan & Hughes 1972), that pigs will rapidly learn to switch on hot and cold air heaters to control their own temperatures (Baldwin & Ingram 1967). Cattle have been shown to respond to an auditory signal to bring them to milking (Albright et al. 1966; Kiley in press). A serious use of such abilities is likely to be beneficial to the animal as well as more economical.

Methods of self-dispensation of food and water, temperature control, movement to the correct place at the correct time, light control in controlled environment houses and a variety of other changes in the environment are open to adaptation for animal operation.

F. Stability of the group
It has been emphasised throughout this section that changing the constitution of the group or individuals in it, whether they are rodents, primates, pigs, horses, cattle, sheep, dogs or fowl, causes disruption and disturbance. For weaner pigs, this is reflected in loss of weight gain (Tindell & Craig 1959), for dairy cattle in drop in milk yield. Other effects which have not been measured to date are a decrease in egg production and weight gain in poultry and an increase in attack in most species (personally observed in horses, cattle, pigs and poultry). Guhl & Allee (1944) and Guhl & Atkeson (1959) found that the longer the

residence of chicks or dairy cattle in the group, the higher the dominance status. Tindell and Craig (loc. cit.) found that moving the piglets around every 28 days reduced weight gain. Many farmers have reported that setbacks are experienced by the piglets when litters are mixed or animals sorted according to size. It is suggested that one way of overcoming this problem is to keep the groups as stable as possible and, in particular, to raise the piglets in stable litter groups from birth to slaughter in so far as this is possible. Some successful farmers employ this method. How much such farms' successes are due to this factor is unknown. This is another area where experimentation unconfounded by other environmental factors would be fruitful.

Recently it has been proposed that dairy cows should be split into groups according to the stage of their gestation or lactation, which would simplify management. However, in view of the fact that a changing group is disruptive and unstable, such management is unlikely to be very successful. Most herdsmen are aware of the individual associations that are formed within the cow herd. Splitting of such associations results in restless behaviour, increase in vocalizations, a decrease in milk production and often in reduced food intake in cattle and other species (e.g. horses, sheep and dogs, personal observation). It seems rather that groups should be as stable as possible, preferably being composed of the same animals from birth to death. Unavoidable short term isolation from the herd (for example, for insemination or parturition, or because of disease) are not likely to have large scale effects on the organization of the group. Brantas (1968a) found that removal of a dairy cow from the herd for several months resulted in no change in social organization but the frequent sorting of animals into different growth stages or phase of lactation is very likely to cause behavioural effects that will reflect on production.

CHAPTER 6

THE EFFECTS OF GROUP SIZE ON BEHAVIOUR

The size of the group has more often been based on the number of
animals one man can look after (e.g. dairy cattle), or the limitations of
the buildings, rather than the demands of the animal. Recently, however,
because of the unexpected drops in production in some intensively
managed animals, particularly pigs, some work has been done on the
effect of group size on production and behaviour. Since the demands of
the different species are very different in this context we will treat each
species separately.

Pigs. It has been shown several times that the size of the group affects
the rate of liveweight gain in pigs (e.g. Wingert & Knodt 1960; Clawson
& Barrick 1962). Heitman et al. (1961) and Gerlbach et al. (1966) varied
both amount of floor space and number of pigs per pen. Heitman et al.
found that in groups of 3 conversion was worse than in the larger groups
(of 12 pigs). This was attributed to larger intake in the smaller group.
On the other hand, groups of 4 - 6 pigs grew faster (conversion rate
unknown) than groups of 8 (Gerlbach et al. 1966) Thus it looks as if
smaller groups of 3 - 6 pigs may do better than larger groups. Indeed,
Fredeen and Jonsson (1957) found that individually penned gilts gained
weight faster than gilts in groups (see, however, isolation, page 38).
Unfortunately, the studies that have been done on group size in pigs have
often been confounded with other variables, consequently there are no
clear conclusions. For example, Gerlbach et al. (loc. cit.) conclude that
the effects of group size vary according to environmental temperature,
type of flooring, temperature variation and amount of space.

Bryant (1971) found that smaller groups of pigs at lower stocking
densities did not have as many agonistic encounters as larger groups. Thus
it would appear that small (say, litter-sized) groups are likely to do best.
However, this clearly needs much further work before new buildings can
safely be designed. I have seen groups of 50 and more weaners apparently
doing well in large barns.

Cattle. There are reports that production in dairy cattle from very large
herds (200 - 300) tends to be less than from the smaller herds. MacMillan
(1970) found that there was a decrease in infertility in dairy herds of
more than 200 animals. However, it has not been shown whether this is
due to less efficient management (i.e. recognition of oestrus) or group
size per se. Mr H. Ellis, Church Farm, Sussex, and many other farmers

I have spoken to insist that single man dairy units (70 - 80 cows) are the most profitable and efficient way to keep the cattle. The Milk Marketing Board have many figures on profitability of herds of different sizes, but these have not to date been analysed.

It would appear that a stable social structure would be more likely to occur where all the individuals recognise and know all the others. This is more likely in a herd of 70 - 80 animals than in a much larger herd.

The Beef Recording Association (1968) showed that Friesian steers in groups of 18 showed a better weight gain than those in groups of 28. Whether or not size of group per individual is the controlling factor remains to be investigated. Thus many cattle farmers prefer to have very much larger groups but give the same space per individual. The argument is that since all animals are not equally distributed through the area all the time it is possible for individuals to move freely through temporarily unoccupied space. In small groups where the allowance of space/individual is the same, this is not the case.

Poultry. This same principle tends to be applied in the broiler industry where groups are very large. However, Mayhew Chickens Ltd subdivide their birds into groups of 2,000 within a floor area holding a total of 15,000 birds. These subdivisions reduce the effects of panicking and smothering and allow slightly more control. Again, there appears to be little work altering only group size in poultry, although Craig (in Guhl 1953) found that, as group size increased from 20 - 100 - 400 chicks, so pecking increased.

CHAPTER 7

THE EFFECTS OF ISOLATION AND SOCIAL FACILITATION

A clear distinction between behavioural effects of isolation and social facilitation is difficult. One of the problems is that many normal behaviours may be inhibited or reduced in isolation in a social animal which is accustomed to being in a group, because the environment is strange and the animal may have physiological responses related to stress (see below). This effect has been termed 'isolation inhibition'. Thus, when assessing social facilitation by comparing the performance of behaviour in isolation and in the group, one may well be measuring 'isolation inhibition', that is, the effects of isolation rather than those of social facilitation (Birke & Clayton in press).

There are various reasons that lead farmers to isolate animals in individual pens. In the first instance, some animals are isolated from conspecifics as they are aggressive and damage others. In indoor pigs, boars are usually isolated from sows for this reason. However, the reason why they are aggressive towards females in the first place is probably because of crowding (see page 15). The isolation of the boar from the sows necessitates more capital expenditure in buildings, increased labour cost and difficulties in identifying oestral sows. Bulls are considered to be 'naturally' very aggressive and largely for this reason are kept isolated from the cows. Calves are often isolated either to facilitate individual feeding or to reduce inter-suckling (see stereotypies). Battery hens are isolated to enable measurement of individual production and to avoid egg breaking.

Animals on the farm may also be isolated for short periods to prevent spread of disease, or for long periods to facilitate ration feeding or other management procedures.

There is little doubt that isolation from conspecifics will affect behaviour in many different ways. However, the extent of this effect will largely be dependent on the degree to which the animal is social in the wild or semi-natural environment, as well as on the animal's early experience. Thus it would be expected that very social animals normally associating in large herds will find such isolation more traumatic than a species that is usually solitary. Little work on the behavioural effects of isolation, unconfounded by effects of confinement or early development, has been done in any group other than rodents. What information there is, together with some of the rodent work, is briefly summarised below.

A. Physiological responses to isolation

Hatch et al. (1965) found specific physiological responses to isolation in rats. In particular they found that adrenocorticotropic hormone (ACTH) secretion increased in isolated animals. They also found that haemoglobin levels increased in both sexes and adrenal, thyroid and pituitary gland weights increased, in females in particular. These responses are similar to those of the General Adaptive Syndrome described by Selye (1950), which results from stress. As a result of these physiological changes Hatch et al. were able to identify an 'isolation syndrome', but showed that it depended markedly on breed, sex and time of the year.

Stein et al. (1960), adding the effects of early experience to those of isolation, found that albino rats isolated early in life withstood the effects of immobilization and fasting better than those that were group reared. The former rats showed less stomach lesions but more 'fearful' (defaecating) responses than the latter. They conclude, rather rashly, from this that a degree of isolation early in life (that is, 'stress') helps towards survival later on.

In any event there is some evidence to show that isolation has a considerable effect on the physiology of the animal, which in many respects could be regarded as a response to stress.

B. The effect of isolation on aggression

The effect of isolation on aggression in mice is so well known in the psychological literature that it is used as a research technique to discover the effect of various drugs and manipulations on 'isolation induced aggression' (e.g. Jansen et al. 1960; Fredericson 1950). Recently Banerjee (1971) in an enquiry into the genesis of this aggression (attack) suggests that it is related to the general hyper-excitement seen in isolated rodents. He suggests this is in turn related to the reduction of invironmental stimuli in such environments, following Welsh's (1964) theory of optimal level of environmental stimulation. A similar increase in aggression has been found in isolated monkeys (e.g. Mason 1960).

Farm animals are also isolated in confined and dull environments (e.g. bulls, veal calves and sows in stalls). It seems probable that isolation may induce or increase attack in farm animals, just as it does in mice and primates. This is obvious in bulls: dairy bulls that are normally isolated from the herd are notoriously dangerous and will attack their own or another species (e.g. man), whereas herds of these bulls can be kept together with few serious aggressive encounters (Hunter & Couttie 1969, personal observation at Whenby Lodge, M.M.B.). Anecdotal evidence suggests that beef bulls kept with the mixed herd are far less 'aggressive' and dangerous than their isolated counterparts. Genetic as well as environmental factors may also play a part in this, since it is generally believed that dairy bulls are more 'aggressive' than beef bulls.

C. The effect of isolation on social and sexual behaviour

Here again the available results are conflicting. Thus Wood-Gush (1958) and Baron & Kish (1962) found no long term effect on sexual behaviour or social behaviour in chicks raised in isolation, in pairs or in a flock. However, Schein & Hale (1959) found a substantial effect on sexual behaviour depending on whether their turkeys were raised in visual isolation from conspecifics or in groups. In the rat, Beach (1942) found that there was an increase in copulation in the isolation-reared males, and that the lowest number of copulations was scored from a group-reared male! On the other hand, Valenstein & Young (1965) found that socially-reared guinea pigs showed more effective copulation that isolates.

Most of the work that has been done on isolation has concerned isolation of young animals and the estimation of this effect later in life. Mal-imprinting and concomitant effects are discussed elsewhere (page 58). The effect of isolating adult animals appears to have received little attention.

D. The effects of isolation on food intake

For the farmer the most universal effect of isolation, as far as we know to to date, is its effect on food intake. There has been little work with farm animals on this effect, most studies assuming that data collected in the laboratory in isolation will be relevant to the group situation on the farm. The work with rodents is conflicting, thus Prychodko & Long (1969) in mice, and Coffey (unpublished) in calves showed that isolation reduced growth, and Harlow (1932) showed that it reduced food intake in rats. Shelley (1965), on the other hand, found that it increased food intake. Hoyenga & Aeshelman (1969) contradict Shelley's result, but recently Cooper & Levine (1973) and Morgan (1973) confirm it by showing increased intake in isolated animals. They suggest that this is because of greater demands for the maintenance of body temperature in the isolate compared with the group-kept animals.

All we can conclude from these studies is that the effects of isolation vary (among horses, for example, anecdotal evidence suggests that some animals eat more when stabled with others, whereas others eat more when in isolation). This could, however, be due to other effects, such as early rearing conditions. There is little doubt that isolation does have some effect and therefore extreme caution should be exercised when attempting to apply data collected from isolated animals to the farm and/ or group situation.

E. Social facilitation

Different workers have had different definitions of social facilitation as Birke & Clayton (in press) point out in a recent review. They adhere to Crowford's (1939) all embracing definition that social facilitation is "behaviour showing increments in the frequency or intensity of responses already learnt by the individual". In this way it can involve imitative behaviour, observational learning (that is, the ability to perform an act after watching another animal perform it), allelomimetic behaviour (or following), infectious behaviour (such as yawning) and social stimulation (for example, synchronization of oestrus in sheep).

However, social facilitation is probably most commonly used to refer to the increase in a particular activity as a result of the performer being with other animals. Clearly this can refer (a) to the increase in intensity with which the activity is performed (e.g. the intensity of running in dogs, Vogel et al. 1950), (b) to an increase in bout length (e.g. Tolman 1968 found that feeding bouts were longer when chicks were fed in groups) or (c) to the initiation of that activity (e.g. Ross & Ross 1949 found that a satiated puppy would recommence eating if placed in the presence of a puppy which was eating).

The physical aspects of the situation are important. For example, when hens are provided with larger piles of food, they eat more (Schreck et al. 1963). This may influence the results in a social environment where the piles of food given to groups may be bigger. The size of the food particles affects intake (page 83) and even the hardness of the substrate in feeding chicks (e.g. Birke & Clayton, loc. cit.).

Social facilitation of feeding is the only aspect of social facilitation that has attracted much attention to date, and even here the picture has lately become somewhat confused. Bayer (1929) showed that the intake of a satiated hen placed with a hungry hen could be increased by 60%, but this may not be only social facilitation since sweeping up the food and re-presenting it induced further feeding.

Tolman (1968) showed that feeding was increased in the presence of other chicks. Ross & Ross (1949) showed a 51% increase in feeding in group-fed pups as compared with isolates. Hafez & Signoret (1969) showed increased intake in pigs in a social environment. Harlow (1932) itemized the characteristics of social facilitation of feeding in the rat as the result of experimentation. He found that it was not subject to change of adaptation, did not depend on imitation, was not due to 'envy' (since the presence of a restrained non-competitive rat did not increase intake) and it did not depend on the size of the group. This comprehensive investigation of social facilitation has, however, not been replicable, and the complexity of the subject is indicated by Shelley's (1965) work where social facilitation occurred when rats were fed once a day.

In the grazing situation Tribe (1950a) divided 50 black-faced ewes into two groups and fed supplements to one group only. When the animals were grouped together their grazing times were very similar, as were their resting times. However, when kept separately the supplementary-fed group spent more time resting and less grazing than the unsupplemented. This he considered to show social facilitation of grazing.

The amount of time performing any other activity may well be subject to social facilitation too, for example, ruminating, lying, suckling and drinking in cattle (e.g. Schmisseur et al. 1966; Kiley, unpublished data) and ponies (Tyler 1968). This may affect production. With these field studies, however, it is extremely difficult to distinguish whether such synchronised behaviour is due to the results of similar circadian rhythms or to social facilitation as defined above. It is likely that social facilitation may act at least to maintain this synchrony.

Social facilitation in the placement of the egg appears to occur in chickens. Thus china eggs are often utilised to encourage egg laying in the right place.

Birke (1974) showed that a remarkable number of behaviours were dependent on the performance of that or another behaviour by a second zebra finch. However, such effects have received no attention in domestic animals to date.

Allelomimetic behaviour can be considered to be a form of social facilitation. It occurs very markedly in sheep (Scott 1945), to a lesser extent in cattle and even less in goats (Scott 1956). It is therefore possible, or even likely, that other behaviours will be more influenced by social facilitation in the sheep than in the other species. Knowledge of such phenomena would be particularly important in the development of the intensively managed sheep industry.

Imitation, another form of social facilitation, has been shown to be important in learning (Scott 1945; Smith 1957; Ross & Ross 1949; and personal observation). This could be of particular use in conditioning animals (e.g. dairy cattle) not to defaecate in the wrong area (e.g. Brantas 1968a and Kilgour & Albright 1971) as well as other forms of conditioning. Animals may also learn by imitation actions detrimental to management (e.g. stereotypies in horses, see page 74, and possible ways through fences, barriers and doors).

It is important that more work should be done on both social facilitation and isolation, since they clearly affect animal husbandry and building design very seriously. Questions should be asked such as: is it necessary to see, hear and smell (or any one or two of these senses) other members of a group, or only one of these, for social facilitation or 'isolation inhibition' to take effect, if it does?

CHAPTER 8
BEHAVIOURAL PROBLEMS ASSOCIATED WITH REPRODUCTION

Another major area of behavioural problems induced by modern management techniques is reduced reproductive performance. The main factors here are:
A. Reduced 'male libido'
B. Difficulties of recognition of oestrus by the handler, particularly in crowded environments
C. Suppression of oestrus as a result of unsuitable environments
D. Deficient maternal behaviour
In this chapter poultry are largely omitted. Recent reviews, such as Wood-Gush (1971) have considered their problems of reproduction which are somewhat different from mammals.

A. Male libido
Because of its importance in the collection of semen for use in AI, male libido has received a certain amount of attention in the last decade. However, there is much disagreement as to how it should be measured. Is it reaction time to intromission (Fraser 1968b), length of time to ejaculation or number of sperms produced per ejaculation (Wiggins et al. 1953), or failure to mount or ejaculate? Others do not mention how they have measured it.

Apart from its use in AI, an understanding of 'male libido' in agricultural animals is becoming increasingly important as a result of the synchronization of oestrus (by photoperiodic control and/or the use of steroids) which results in periodic over-use of males, alternating with periods of no use. This is of particular importance in rams. Tomkins & Bryant (1973), for example, found that conception decreased with oestrally synchronised ewes from 86% (with 6 ewes) to 56% (with 30 ewes). In addition, the semen depleted very rapidly over the first 20 to 30 ejaculates.

The main problems of the collection of semen for artificial insemination are:
(i) Teaching the male to mount the non-oestral female or simulated female (this may be anything from a real female to a padded box).
(ii) Habituation to the simulated female and the surroundings.

These factors result in a reduction in the amount of semen collected and even sperm mobility within the semen.

Willingness to mate is seasonal in some male ungulates (e.g. rams, Bishop & Walton 1960) and there is some evidence to indicate that spermatogenesis is seasonal in more species than has usually been assumed (e.g. Skinner 1972). What controls the seasonality in mating is largely unknown. It has been suggested that better nutrition may allow it to continue through the year. However, Reed (1969) stated that it does not appear to be affected by a poor diet in boars. Bane (1963) shows that underfeeding in cattle does not affect mating and the production of sperm, although overfeeding can reduce it. Similarly, Fraser (1959) listed obesity as one of the reducing factors of 'male libido' (sexual behaviour) in the bull.

Photoperiodicity and temperature are probably important (cf. Yeates 1965 in this context). As with females (see control of oestrus), social factors are also likely to be important, for example, the presence or absence of oestral females, and even other males. However, the problem is not so simple since the presence of too many oestral females may reduce mating in the ram; this happens where ewes are brought into oestrus in synchrony. Under such conditions the ratio of rams to ewes must be increased so as to ensure conception.

Bulls and boars do not appear to show seasonal fluctuations in mating. However, mares generally do show seasonal oestrus and it is often, in consequence, assumed that stallions should also show seasonal changes in behaviour, but this has not been demonstrated to my knowledge.

The next point is that there are very considerable individual differences in the amount of sexual behaviour exhibited by genotypically different males. Bane (1963) showed that bulls on the same feeding regimes showed a greater difference in libido than identical twins on different feeding regimes. Fraser (1959) lists seven different factors that can affect sexual behaviour in the bull, including breed differences.

The dangers of 'sexual frustration' for increasing aggression (attack) and decreasing the production of mobile sperm in the male are often assumed. There is no experimental evidence that sexual frustration increases attack, although wild mammal studies suggest that this is so since, for example, fights between territorial males occur most frequently in the presence of oestral females. That sexual frustration reduces the performance of sexual behaviour is equally untested. Nevertheless, cocks low in dominance ranking show less sexual behaviour and mounting of females even when the dominant animal has been removed (Guhl et al. 1945). Such an animal was termed a psychological castrate. No tests on any measures of sexual behaviour in relation to any measure of social position or dominance of cattle, horses or sheep have been conducted to my knowledge.

Miller (1950) suggested that fighting increased 'libido' (it is not clear how he measured this) in the bull, since the best results were achieved from the most 'aggressive' animals. Particularly when using animals for semen collection, it is suggested that pre-copulatory stimuli are important; thus pre-copulatory stimulation of the male has been shown to increase libido and sperm production (Kerruish 1955; Hawker 1963, in bulls).

Barker (1960) considered that visual stimuli were the most important in the bull, whilst Hart et al. (1946), as a result of non-impairment of mounting by blindfolding, held that olfactory cues were most important. In the ram, gross changes in the appearance of the ewe and partial change in olfactory stimuli did not reduce libido (Beamer et al. 1969). It is probable that all the various cues are important in libido. Nevertheless, the male will perform sexually in the absence of one modality (e.g. blindfolded horses, bulls and rams, Wiezbowski 1964; anosmic rams, Lindsay 1965). Lindsay found that anosmic rams compensated for their lack of olfactory stimulus by increasing receptiveness to auditory and tactile stimuli from the female. Although sexual performance was not affected, the anosmic rams had difficulty in distinguishing oestral from non-oestral females.

Sambraus (1971) stresses the importance of pre-copulatory stimulus of the bull for semen collection, but concludes that the attraction of the 'teaser' object (oestral female or simulated female) is not so important, although it may induce some animals to copulate when they would not do so with a more alien stimulus.

Sambraus (loc. cit.) showed experimentally that the optimum height of the 'teaser' object was 124 cm. but it was immaterial whether it was living or inanimate once the males were experienced.

The last finding leads us to the importance of learning for ejaculation to occur in a semen-collecting situation. An oestral female normally has to be used to induce mounting and ejaculation in the young bulls and rams. With training these animals will perform with crude models of females, although the rams will eventually also mount castrated males in the collecting arena (Lindsay 1965) and bulls other bulls. Nevertheless, olfactory stimuli from the female have been shown to increase production of sperm in the boar and other animals. On the other hand, Trimberger (1962) showed that there was an eventual decrease in libido (measured by time to ejaculation) in bulls used for AI and he found that if a new teaser was presented, or the location changed, the production of semen increased.

There is little doubt that changing the female can re-induce 'libido' in fatigued or satiated animals (e.g. bulls, Milovanov & Smirnov-Ugrujumov 1940; in boars, Pepelko & Clegg 1965). Beamer et al. (1969) showed that the introduction of a new oestral ewe to rams that had copulated to satiation re-induced copulation, and that the initial ejaculatory interval was the same with each successive female. The same group (Beamer et al. 1969) showed that the successive introduction of different females after each ejaculation can maintain rapid copulatory behaviour in some rams through at least twelve ejaculations. Such an effect is well documented in rodents, and is known as the 'Coolridge effect'.

Hale & Almquist (1965) and Fraser (1968a) showed that changing the place where mounting took place re-induced sexual behaviour in bulls and rams.

The effect of isolation on male sexual behaviour should be investigated. It has been suggested that it increases libido, or that it decreases it, but no experimental evidence has been given. In primates the effects of isolation on sexual behaviour are various (Mason 1960). When isolated young, the males usually show reduced and wrongly orientated sexual behaviour as do jungle fowl (Kruijt 1964). Nevertheless, there appears to be no clear experimental work investigating the effect of isolation on adult males isolated as adults, or those groups of males kept in isolation from females. These are the most common conditions under which bulls and boars are kept today and such an investigation would be worthwhile. Hafez (1951) showed breed differences in rams' sexual behaviour but also that these were considerably modified by the behaviour of the ewe.

Thus, although mating may be increased by selective breeding, it is evident that exogenous stimuli are of great importance in the maintenance of it. A greater comprehension of the relative importance of exogenous stimuli in conjunction with breeding selection would be particularly useful. Initially, however, a clear statement of what is meant by 'male libido' should be made by workers in the field, since it is possible some treatments might affect one measure (e.g. time to intromission) while not affecting another (e.g. sperm count per ejaculate).

B. Recognition of oestrus

With the large scale use of AI in cattle and its future extension to pigs and sheep, the ability of the handler to recognise oestrus in these species becomes of particular importance. Where the animals are isolated or in very large herds, it becomes more difficult (e.g. Esslemont & Bryant 1974). Although there are some physical signs of oestrus in most species (e.g. the swelling of the vulval region and secretions from the vulva), the main indices of oestrus are behavioural. Let us examine what these indices are in the various species.

Cattle. Fraser (1968b) lists 9 signs of oestrus in the cow. These are:
(1) Excitement (presumably involving an increase in locomotion and other activities.
(2) Bellowing: vocalization is said to increase during oestrus.
(3) Licking other animals.
(4) Mounting other animals.
(5) Standing to be mounted. Mylrea & Beilharz (1964) show that this is a better measure of oestrus than mounting other cows.
(6) Jerky movement of the lumbo-sacral region.
(7) Clear mucus from the vulva.
(8) Loss of appetite.
(9) Drop in milk production.

Individual animals of one species show different intensities of heat. Low intensity heats or 'silent heats' can be impossible to diagnose in the absence of the bull.

Post-partum heat is frequently silent (that is, there are no behavioural indices of oestrus), although ovulation does apparently take place and the animal is fertile. Thus, unless the individual animals in a herd are watched and the dates on which they are expected to come into heat known, oestrus is missed.

Sheep. AI is not a common practice in sheep. The ram is left to diagnose oestrus in the ewe. There is very little information concerning the detection of oestrus in the ewe, other than that she seeks out the ram when in oestrus and shows 'restless behaviour' in pro-oestrus. Tail wagging apparently occurs in the oestral sheep (Fraser 1968b).

For the successful application of AI in sheep, more information must be accumulated on the oestral behaviour of these animals with a view to identifying possible behavioural indices. One of the alternatives is to keep vasectomised rams in the ewe flocks to diagnose oestrus.

Pigs. The detection of oestrus in pigs is again usually left to the boar. However, AI is being increasingly used in pigs, and there has always been the problem of diagnosing oestrus in the small unit where a boar is uneconomic. Signs of oestrus are a swollen reddened vulva, sometimes accompanied by a mucus secretion. A behavioural index of oestrus is the receptive stance, when the animal will stand immobile often for several minutes when pressed on the back or ridden astride by a man. During oestrus gilts or sows will occasionally mount each other, but this is rarer than in cattle. Sows will also respond to the boars' 'chant de coeur' vocalization (Signoret et al. 1960) by standing. The smell of the boar or preputial fluid has also been shown to induce standing in the oestral female (Dutt et al. 1959). Salivary gland ablation in the boar reduced standing in sows, and the pheromone responsible for this has been isolated (Perry et al. 1972). In addition, breeds of pigs with erect ears are said to show backwards pricking of the ears (Fraser 1968b).

Detection of oestrus is difficult in the sow, since these indices are not always obvious. However, I have observed that male dogs often show an interest in oestral sows, mounting them and making thrusting movements. It is possible that dogs might be trained to be useful indicators of oestrus in sows. Previous experience with dogs would, of course, have to be given to the sows.

Horses. Oestrus in the horse is recognized by a characteristic frequent lifting and sideways displacement of the tail (Fraser 1968b). Urination is more frequent and the performance is affected by the animal behaving unreliably under saddle. It is easy to detect in the presence of other horses, even castrated males and other mares, but more difficult in the isolated animal.

Mares may remain in season for as long as 8 days (Marshall 1965) during which time they may be receptive to the stallion and, indeed, may solicit him by rubbing, licking, vocalizing and turning their rumps to him. They are not, however, fertile throughout the period. A young stallion may consequently be too fatigued by previous copulations to ejaculate effectively during the relatively short fertile phase of oestrus (personal observation). This can be a problem where stallions are run with mares. With experience, however, the stallions learn to ignore the initial soliciting by mares (Wierzbowski, personal comment).

Goats. Fraser (1961) notes that pro-oestrus is recognised in this species by restlessness, by bleating and rapid tail wagging. In the presence of the male, alignment of the male behind the female as opposed to at an angle denotes the onset of oestrus.

Recognition of oestrus by man. For management reasons, even if a male is present, it is necessary to know when females have been in oestrus in order to:
(i) remove anoestral animals
(ii) be able to calculate parturition dates.
This is frequently given as the reason for separating males from females.

However, the recognition of oestrus in a female is greatly facilitated by the presence of a male in all species. The problem of recognising inseminated ewes has been overcome by the use of a coloured chalk as a harness carried by the ram which rubs off, marking but not staining the wool of the ewe's back (Radford et al. 1960).

A comparable apparatus is available for bulls involving a chin ball with ink in it held by a head collar. The ink rubs off on the rump of the tended cow during chin resting (Yeates & Crowley 1961). A similar harness with a marker could easily be worn by the boar and stallion, which would allow boars and stallions to be kept with the females whenever this is desirable for other reasons.

C. Disturbances and suppression of oestrus

(i) The effect of light on oestrus. By manipulating light regimes, 50 to 60 days in the dark followed by 11 hours light, ewes can be brought into oestrus in synchrony and out of season (Mitig 1968). This allows breeding twice in one year and increase in production.

Although cattle are supposed not to show seasonal breeding, Deas (1970) found that an increase in light intensity in the cow-shed decreased the length of time between parturition and conception. This could be the result of increased fertility or better recognition of oestrus by the stockman with increased light. The control of oestrus in other mammals by the use of photoperiodicity is an obvious development. While suggesting the reduction of light intensity in modern buildings such as chicken and pig houses (to reduce aggression and cannibalism), the effect this may well have on oestrus and egg laying must be considered. Reduced lighting in sow houses is frequent and one of the main problems, particularly in stalled animals, is the recognition of oestrus and conception.

(ii) Genital stimulation of the female to induce oestrus. Van Demark & Hays (1952) and Hays & Carlevaro (1959) found that genital stimulation of the cow either electrically or by hand increased the conception of the females at oestrus. It is suggested by these authors that the mechanism by which this operates is to increase the rate of sperm transport. Pre-copulatory courtship of the female by the male involves much nudging and licking of the vulva in most mammalian species. This results in various secretions into the vulva which also possibly increase the rate of transport of sperm, thereby increasing conception.

(iii) Presence of a male. There is some evidence to suggest that the presence of a male increases the fertility of females. Whitten (1956), confirmed by Marsden & Bronson (1964), found that oestrus was synchronised and induced in mice by the introduction of male faeces or urine. Coleman (1950) and others have since recognised this effect in

sheep. Schinckel (1954) reported that the presence of a ram stimulated reproductive activities in ewes, and Bellinger & Mendel (1974) found this to be the case even outside the breeding season. Fraser (1968a) showed that lambing occurred at a much faster rate and began and finished sooner in a group of Suffolk sheep that had been with a vasectomised ram for five full weeks before the beginning of the tupping (mating) season, compared with a control flock that had no pre-tupping contact with a male.

Polikarpova (1960) observed that sows kept with boars had more fertile heats than sows kept apart. Fraser (1968b) found that oestrus in anoestral gilts was induced by the introduction of a boar to the premises within 48 to 72 hours.

Petropavlovskii & Rykova (1958) ran a teaser (a vasectomised bull in this case) with a herd of freshly calved cows for 3 - 4 hours daily, while control cows were kept away from the bull. The teased cows bred 4 weeks earlier than the control and showed "more marked oestrus" (that is, oestrus was easier to identify). Nersesjan (1959) found that teased cows come into heat earlier than the controls.

Even litter size is affected by the presence of the male in mice (Beilharz 1968), although a similar effect has not been reported on twinning in sheep to date.

Van der Lee-Boot (1956) showed that groups consisting of only female mice showed oestral suppression, became pseudo-pregnant or anoestral; and recent work with women, which showed that the presence of males has a shortening effect on the oestral cycle (McClintock 1971), would suggest that fertility may be reduced in the absence of a male by suppression of oestrus.

It has often been assumed that the female will stand for the male willy-nilly when in oestrus, but studies of wild ungulates mating (e.g. Walther 1974) suggests that the standing of the female is only induced after prolonged courtship and stimulation by the male. That females show preferences for particular males is well known by dog breeders. The soliciting of mating by the female has only recently been considered important in mating. Fletcher & Lindsay (1968), for example, point out that sight, smell and hearing are important in ram-seeking behaviour in oestral ewes.

Whitten (loc. cit.) showed the importance of pheromones in mating mice and further evidence of their involvement in pig mating comes from Dutt et al.'s work (1959). Recent work by Perry et al. (1972) showed the importance of the sub-maxillary salivary glands in mating. The substance responsible was an androsterone. It is possible that an abstract of the substance could be used to spray on sows to detect oestrus where boars are not available. However, since visual and auditory cues from the boar are likely to be important to the sow it may prove more economic, particularly in larger units, to keep either an entire or vasectomised boar with empty sows.

(iv) Other behavioural effects of oestrus. Disruptive behaviour and an increase in activity are typical of oestrus (see above). Where animals are close-herded, this can cause very real problems. In dairy cattle kept in a yard, the disruptive effect of one oestral animal is enough to cause a noticeable drop in the herd's milk yield (P. Savage, Colin Godmans Farm, Sussex). In addition, mounting each other, moving and pushing each other (typical of oestral cows) can cause considerable damage to confined cattle (e.g. breaking of legs, hips dislocated, teats stepped on and damaged). This is a more extreme problem in animals confined to small areas than in the open field grazing situation, where there are ample opportunities to avoid the disruptive animal.

The presence of a bull, either in the herd or in an adjacent yard, might help to reduce such problems, as the oestral animal will be tended by him and keep close to him, reducing herd participation.

A similar problem is found occasionally among gilts and sows confined to stalls or in yards. However, even with the male in the same shed, problems of oestral recognition arise.

(v) Suspension of oestrus. It is well known that very poor physical condition or emotional shock can lead to a suspension of oestrus cycles in women. Poor physical condition due to starvation is uncommon in agricultural animals in Britain today but severe physical stress, such as rainfall and wind without shelter, are encountered by hill sheep, for example. Such conditions have been found to suppress oestrus and mean ovulation rate in ewes (Doney et al. 1973). Whether this is due to suspension of oestrus or to foetal reabsorption, still-births or neo-natal deaths has not been shown to date.

Evidence is accumulating to show that suckling delays the onset of oestrus. Wiltbank & Cook (1958) and D'Alba (1960) showed that cows milked twice a day conceived more quickly post-partum than their suckled counterparts. Single and double suckled beef cows are slower to start their oestrus cycles post-partum than dairy cows (M. Kiley, unpublished data). Pigs do not as a rule come into oestrus until 4 to 10 days after weaning, depending on when they were weaned; if this is earlier, then oestrus is later.

Warnick et al. (1950) found that the follicular volume of suckled sows was significantly smaller than the non-suckled sows. Also, animals milked 3 to 4 times a day often have delayed post-partum oestrus. On the other hand, although the number of times milked or suckled and therefore the release of oxytocin may be important in controlling follicular growth, the presence of the young may have some considerable effect on the delay of oestrus. Peters et al. (1969) showed that, whether pigs were mammectomised or not, it was the presence of the piglets that decreased the release of follicular stimulating hormone. There is anecdotal evidence to suggest the presence of young, as well as more frequent milk let down, delays oestrus in beef cattle.

This effect of the presence of the young could be confounded in cattle by the 'stockman effect' (see section on stockmanship) where a good stockman may be considered to be in loco parentis, thus shortening the period to milk let down, but also affecting fertility by his presence.

D. Deficient maternal behaviour

Maternal behaviour, its triggers and controls is one of the aspects of behaviour we know least about. There has been some work done on its physiology, mainly in the rat (reviewed by Lehrman 1961 and Rosenblatt & Lehrman 1963). Maternal behaviour appears to be to some extent under under hormonal control. Thus Ross et al. (1963) showed that ovariectomised rabbits treated with oestrogen, prolactin and progesterone could be induced to build nests and even lactate. Lehrman (loc. cit.) emphasises the importance of external stimuli triggering maternal behaviour. Rosenblatt & Lehrman (1963) concluded from experiments with rats (in which they maintained the maternal condition for long periods by exposing the mother to young pups which were replaced periodically) that the stimulus from the young is very important for the maintenance of the maternal condition. Thus the two most important factors in controlling maternal behaviour appear to be the hormonal state of the mother and external stimuli from the young.

(i) Baby stealing. Most ungulate parturient mothers tend to isolate themselves to a varying degree from the rest of the herd (Fraser 1968b; Lent 1974). Clearly the effect of this is to help the mutual imprinting of mother and young, essential for survival thereafter. If, however, parturient animals are kept in confined areas, then there is no opportunity for this and mis-mothering may result (see below). In addition, maternal behaviour appears to be triggered off in some females before actual parturition. These pre-parturient individuals may try to mother other young and, indeed, steal them away from their mothers before they are mutually imprinted. This is particularly a problem in sheep which are often confined at lambing. In dairy cattle the mass mothering of calves calved in the herd may result in death from trampling. One farmer in Sussex loses 25% of his calves in this way (personal comment Mr P. Clarke, M.R.C.V.S.). Such behaviour also occurs in single-suckler beef cattle and sheep at pasture (Kiley, unpublished observation). On the other hand, isolation of the pre-parturient females has its problems, one being the development of over-solicitous behaviour towards the young.

(ii) Over-solicitous maternal behaviour. If mother and young are isolated together for long periods of time, the mother may develop abnormal behaviour of one sort or another towards the young. One of these is over-solicitous behaviour. She may, for example, lick the young too much and too long until it becomes devoid of hair. This is more a problem in zoos. There is little reason why mothers should be isolated for prolonged periods on the farm, but it is a type of behavioural development that should be borne in mind when designing buildings or developing new husbandry methods.

(iii) Recognition of approaching parturition. It is important to be able to recognise imminent parturition in order to isolate females. This will prevent baby stealing or interference from herd members or, indeed, enable close observation for veterinary purposes. There are physical signs, such as swelling of the vulva and filling of the mammary glands, but these may take place well in advance of parturition (Arthur 1961 and 1965). Thus behavioural changes often have to be relied on, although they have not been systematically studied. Arthur states that mares, ewes and cows become more restless, ruminating is frequently interrupted and vocalizations of low amplitude may occur. Most ungulates withdraw from the herd or group immediately before parturition but the degree to which this is obvious depends on the species (Lent 1974) and on having sufficient space to withdraw in!

Recognition of imminent parturition is especially difficult in the mare, in which it is particularly important, since complications in labour are common in thoroughbreds. Mares apparently have the facility to interrupt labour if conditions are not suitable, just as has been reported for the gnu (Estes 1966). Studies of behavioural changes at this time are needed. The sow, like the rabbit (Rose et al. 1963) and the rat (Lehrman 1961), shows pre-parturient maternal behaviour in the form of nest building, which is a useful index. However, under modern conditions the sows usually have no opportunity to show such behaviour. Parturition can be induced by drugs such as dexamethozane in ewes.

(iv) Mis-mothering. This is the term used for any abnormality in maternal behaviour. It normally takes the form of rejection of the young, but may also involve attack and even eating the young (in rabbits and pigs, for example). Why this behaviour occurs in some animals and not others is obscure. 'Nervous' or very reactive sows appear to be more prone to behave in this way than more tranquil ones (Dr M. Peason and personal observation). Kristjansson (1957) had some success in using tranquillizers to prevent sows eating their young.

(v) Lactation. The importance of the sucking reflex in causing the secretion of oxytocin has long been known (see Turner 1961). In dairy cows the sucking stimulus of the calf has been replaced by the milking machine. However, it is well known that emotional disturbance can inhibit the release of oxytocin and so milk let down (Cross 1955). In addition, milk let down may be possible to condition; thus it is suggested that the noise of the buckets, sight of the milking shed and specific milkmaid may act as important conditioned stimuli for milk let down (e.g. Baryshnikov & Kokovina 1959).

It has been found that routine practices and permanent staff in the milking shed have their worth reflected in the increased milk yield. Since milk let down can be conditioned, it might be useful to condition the cows to let their milk down to a specific stimulus (see stockmanship).

(vi) Mutual imprinting. The initial recognition of the young by the mother is the crux to mutual imprinting, and better understanding of the establishment of mother/infant recognition in agricultural species is imperative if mis-mothering and rejection of the young are to be avoided.

There is evidence to show that in sheep, goats and pigs, at any rate, there is a critical or 'sensitive' period a few hours after birth during which the mother learns to recognise her young (Blauvelt 1954; Klopfer 1961; Hersher et al. 1963).

Many young (lambs, foals, calves and piglets) are willing to suck from any mother that will let them (see adoption, below). They will even attempt to suck from inanimate objects (personal observation).

Thus the problem of recognition lies, in this context, mainly with the behaviour of the mother and her recognition and tolerance of the young. Non-recognition of the young in mother-reared animals (e.g. ewes, beef cattle, sows and horses) will clearly result in economic disaster. This is a considerable problem in close folded animals such as sheep at lambing time. These animals are being bred to produce two or three or even four young. The pressure of numbers must add to the difficulty which the mother finds in learning which are her young in a crowded environment. In addition, if we knew more about the controls of mutual imprinting, we could use it to our advantage to make one mother adopt extra young (for example, in beef cattle).

The modalities available for use in imprinting are, of course, olfaction, sight and hearing.

Olfactory cues appear to be of particular importance in the recognition of a young by the mother. This has been known for a very long time. The traditional method of getting another mother to adopt an orphaned foal, lamb, kid or calf was to disguise its smell with the amniotic fluids from the mother (Collias 1956 even attracted a ewe with an object covered in birth fluid), the skin of her young (e.g. Rossdale 1970), or the milk of the mother. Provided this is done sufficiently soon after parturition, the mother often accepts the foster young. More recently, Crowley & Darby (1970) have shown that masking the sense of smell by smearing the nose of the cow with linseed oil can increase the chances of her tolerating other young. Olfactory stimuli are not the only cues used even within this critical period of the development of attachments: Collias (loc. cit.) showed that washing the lambs in detergent did *not* prevent them from being accepted by their mothers. Baldwin & Shillito (1974) confirmed this when they found that anosmic ewes could form normal bonds with their lambs. Kiley (1976), using a nasal anaesthetizer, had a similar result in beef cattle.

Visual cues are also important. Rossdale (1970), for example, describes how a mare rejected her own foal when its appearance was changed by the foal wearing a head collar. Murie (1944) found that lambs could not recognise their mothers after they had been sheared. Crowley & Darby (1970) found that the likelihood of adoption was improved when cows were blindfolded. Thus visual cues do appear to be of some importance in mutual recognition of mother and young, and it is possible that visual cues may be used more as the young matures (Kiley 1976). Hersher et al. (1963) found that lambs raised by goats and vice versa would recognise their foster mother from a distance very rapidly.

Mother-young contact calls are common throughout ungulates, canids and felids (Kiley 1972). Apart from merely identifying the vocalizer as a particular class of animal, it is possible that mothers and young recognise each other's voices. However, there is no experimental evidence in cattle or other domestic animals to indicate this, although Espmark (1971) has shown that there is recognition of their own fawn's call by the mother reindeer.

It is evident that we know very little about how a mother learns to recognise her own young, let alone what changes arise as the animal matures. Nevertheless, apart from the problems such as non-recognition of the young that we have already mentioned, there are two other important areas in which application of mutual imprinting is used in agriculture. These are:

(vii) Adoption. It has already been mentioned that the young will tend to be willing to suckle any lactating female in the group if the female will let them. However, Hersher et al. (loc. cit.) took this one step further and showed that, provided the mother was confined, sheep could be fostered onto goats and vice versa, although the length of time for acceptance of the young by the mother would vary. The animals were then released with the herd and there were no reversals of this adoption. This illustrated that in enforced isolation the mother will learn to accept young, even of the wrong species, as her own. The importance of the age of introduction to the female and the length of time following parturition for success in this process is unknown. Other variables may also be important: Blauvelt (1954) found that if the foster young were substituted for the ewe's own, then they were accepted, but if they were added to the ewe's own young, then they were rejected. This was confirmed by Kiley (1976) in cattle. This suggests that either sheep can count up to two, or that the facility to make continual comparisons between the two young enable easy identification of her own. The attitude of the mother to introduced calves will often change with time. Normally the mother becomes more tolerant to her own adoptees, as well as others (Kiley, loc. cit.).

In addition, there is perhaps something in the actual process of giving birth which triggers off maternal behaviour, since cows delivered by caesarian section often will not accept their calves until confined with them in isolation for some days (personal comment, Mr Cox, M.R.C.V.S.).

(viii) Fostering. This final category is made use of in multi-suckler systems of beef or pig rearing, but not always with success due to the little knowledge available in general on maternal behaviour.

Some cows will be more permissive than others. Kilgour (1969) found up to eight calves suckling one cow in a multi-suckler herd, while another cow would suckle only one. The result of this is that some calves obtain a lot more milk than others. In multi-suckler systems one would like the mothers to adopt the calf or calves and look after them all as if they were their own, as opposed to tolerating them either more or less. Many owners of multi-suckler herds find that, although they can get the cows to allow suckling, the difference in growth rate between her own and her fostered calves is very marked in her own calf's favour. Hersher et al. (loc. cit.) suggest that 'dominance status' is important in whether the female will foster young or not. In so far as this may reflect general personality traits it could be so. However, a closer understanding of dominance is essential before this will be clear (see page 18).

(ix) The effect of mother rearing. If the animal is mother-reared (as opposed to mother-isolated) there is an effect seen in the eventual behaviour of their young, and what type of a mother they become. Zarrow & Denenberg (1964) found that rats and rabbits reared in isolation showed defective care of their young; this was also shown by Mason (1965) with rhesus monkeys. However, recent work with heifer dairy calves (Donaldson et al. 1972) indicates that isolated-reared animals were better mothers than group-reared, although there were no mother-reared calves to compare with. It may be that the experience of having a mother is important for the proper development of maternal behaviour. Clearly it is important to know in domestic animals what effect different types of rearing will have on the maternal behaviour, when the animals are being reared to be mothers themselves.

(x) Effects of artificial rearing. Liddell (1956) found that isolated sheep were less responsive to conditioning and more lethargic than their mother-reared twins. Lemmon & Patterson (1964) found that mothered lambs learned a conditioned response more quickly and made avoidance behavioural adjustments better than the unmothered animals. Thus it looks as though the experience of having a mother can have quite a profound effect on conditioning and response to 'stressful' conditions. This should be borne in mind when advocating further mother-isolated units of commercial animals, as behavioural differences or abnormalities may be seen in such animals.

Kilgour (1969) suggests that dairy cattle should be bucket-reared and handled, since they will have to be handled when adult. I would dispute this on the grounds that they will have ample time to learn the dairy routine later and adverse behavioural symptoms may result from bucket-rearing. Horses, whose whole commercial use lies in their ability to be conditioned and handled, are often reared away from humans and are unhandled and wild at 1 to 2 years but, nevertheless, are trained successfully thereafter.

There are often difficulties in teaching calves to drink from a bucket. Dairy calves usually learn more easily than beef animals (personal comment farmers and personal experience). In some instances beef animals have to be rejected from calf-rearing units because they will not learn to drink. At present calves are usually taught to drink by allowing the calf to suck a finger and then lowering the hand into the milk in the bucket. Various difficulties are encountered in such a method. The hand is frequently bitten and a great deal of time has to be spent on each individual for 2 - 7 days. Various gadgets have been tried to help the calf learn but most farmers find them ineffective. The provision of teats on buckets is usually not adopted as capital expenditure and labour in cleaning is increased. The main learning problem for the calf in such a situation is to lower its head. Searching for the teat is performed from shortly after birth with the nostril out and head tilted upwards. Thus gadgets allowing in the first instance this stance to be maintained would most likely be successful. However, teaching a calf to drink from a bucket is far easier in the first week of life than after normal suckling has been continued for several weeks (Kiley, personal experience).

If severe behavioural problems are to be avoided in raising isolated young animals, than attention must be given to the social relations within the group; these may control food intake in close-folded animals as well as other forms of behaviour (see e.g. Intake, page 80 et seq.).

An investigation is also needed of the effect on growth rates and behaviour when piglets are raised on machine mothers. Such raising methods may become important in the future, as the sows would then be able to produce more litters in a year.

Multi-suckling methods in which two to six sows are kept together with their piglets appear to be successful. However, details are needed on (1) the amount of cross suckling, (2) differences in permissiveness of the mothers and (3) behavioural changes as compared with more conventionally managed single sow and piglet units, not only for academic interest but also for the selection of sows for such management. Further detailed work on double suckling of beef cattle would also lead to increased production.

CHAPTER 9
CASTRATION

Castration is practised on most male farm animals in Britain. There appear to be four major reasons quoted for this, viz.:
A. Castration reduces male aggression and thereby makes males easier to handle.
B. There is a risk of unwanted matings if there are uncastrated males on the farm, other than those required for breeding.
C. Due apparently to traditional butchers' prejudices, uncastrated male meat is marketed cheaper than that of castrates.
D. Performance and conversion rates can be improved by castration.
   There are, however, disadvantages to castration. These are:
(i) The cheaper methods of castration, which the farmer can do himself, require a degree of skill. With such methods there is a risk of infection.
(ii) The alternative, of calling in a veterinary surgeon, adds to costs.
(iii) Castration involves some time.
(iv) It is a traumatic experience, probably more so with older animals. Although there are laws demanding anaesthesia for castration of animals over a certain age, this often involves the animal being herded, isolated and handled for the first time, which may well be considered a traumatic experience for the animal.
(v) There is normally a loss in weight gain and viability for a period following castration in the young animal. This is particularly obvious in piglets and calves, although the weight loss may only be small if castration is correctly performed.
   Although there have to date been several reviews on the effects on production of castration (e.g. Turton 1962; Robertson 1966) there does not appear to be any work which covers all aspects of castration and which in particular considers behavioural questions. The points from the introduction are therefore discussed in the following review.

A. Factors influencing male aggression
The effect of castration is normally to decrease the production of testosterone. This hormone is usually considered responsible for male-like behaviour and an increase in aggression in males. Most evidence for this assertion comes from laboratory animal studies (Archer 1974).
   Environmental conditions can also be significant in inducing aggression although they have not often been considered in the argument to date.

An increase in aggression in confined farm livestock is common (see page 8 et seq.). Whether this is due to crowding or to other physical and dietary components of the environment is not known.

Isolation of male animals is also common practice in present-day management systems and this, too, can have an effect on the level of aggression as shown from studies of laboratory mice and monkeys (Mason 1960). In this case it has been suggested that isolation leads to a reduction in the general level of environmental stimulation with consequent hyper-excitability. This in turn is related to the genesis of aggression (Banerjee 1971). Bulls, stallions and boars are often kept alone in impoverished surroundings and it seems probable, although not yet confirmed, that such isolation leads to increased aggression.

There is some evidence that dairy bulls which are usually isolated from the group in pens are aggressive and dangerous, whereas herds of dairy bulls can be kept together with few serious aggressive encounters (Hunter & Couttie 1969 and personal observations). Beef bulls normally kept with the herd and treated as one of the herd are far less aggressive and dangerous. However, there is a need for more controlled behavioural studies of isolated and herd-kept bulls.

The way a farm animal behaves and hence the level of aggression it shows will also be greatly influenced by the way it is handled by the stockman, particularly when young. Bulls, boars and stallions are normally considered difficult and aggressive to handle. Lack of confidence or fear on the part of the stockman may actually induce more agressive behaviour on the part of the animal than might otherwise be the case. A complex form of dominance/subordinance relationship exists between a stockman and the animals in his charge so it is apparent this is an important factor affecting levels of aggression.

Other behavioural effects of testosterone. Testosterone is not only involved in aggressive behaviour. It is likely, though not yet well established, that testosterone controls many glandular secretions in ungulates. For example, it appears that testosterone may control release of the male rutting odour (Fletcher & Short 1974).

The production of specific odours by the male is, for instance, important in the pig, where the smell of preputial fluid is involved in eliciting a standing reaction from the female (Signoret & du Mesnil 1961).

The effect on females of the presence of entire males. Another behavioural effect of having entire males on the farm is its effect on the females. There is a positive advantage here in that the presence of the male increases the fertility of females and can synchronize oestrus. This was shown by Coleman (1950) in sheep. Fraser (1968a) showed that lambing occurred at a much faster rate and began and finished sooner in a group of Suffolk sheep that had been with a vasectomized ram for five full weeks before the beginning of tupping, than the control flock which had no pre-tupping experience with the ram. This is not only true of sheep as Polikarpova (1960) observed that sows kept with boars had more fertile heats than sows kept apart. Fraser (1968a) found that oestrus in anoestral gilts was induced with 38 - 72 hours after the introduction of a boar to the premises. Similarly, the presence of vasectomized bulls has been shown to affect oestrus and breeding in cattle (Petropavlovskii & Rykova 1958; Nersesjan 1959).

B. The risk of unwanted breeding by non-castrates

Routine castration of males on the farm, except those required for breeding, began when most animals were kept at pasture or free-range and therefore the risk of breeding from unwanted males was high. At the time this practice clearly had a very advantageous effect on animal breeding. Today, however, there is a growing necessity to house more animals and the risk of unwanted breeding is reduced. Separate housing is required, however, for males and females if they are to be marketed after puberty. Such a disadvantage must be offset against the production advantages of entire males (see below). Today, with more separate utility designed units being built daily, separate housing of males and females is not so economically demanding as on the more traditional farm.

C. Butchers' prejudices

One very important factor which ensures castration in pigs, sheep and cattle is that butchers on the whole have a prejudice against buying entire male carcasses. As Clark (1965) points out, the carcass quality of the rams is often rated lower simply because it is ram rather than because of any inferior quality. This prejudice appears to have no factual basis (with the exception of the tainting of boar meat). There is no evidence to suggest that bull or ram meat or pre-pubertal boar meat is in any way inferior to that of the castrate. Nevertheless, the prejudice against entire male carcasses is now such that many farmers believe it to be illegal to raise uncastrated pigs or sheep. An experimental litter of 12 piglets were raised to 9 weeks by the author. Although well grown, the weaner pool refused to buy the pigs because of the presence of 7 uncastrated males. Eventually they had to be sold at a price approximately 40% below the market value.

D. Behavioural and performance effects of castration in the individual species

Cattle. The importance of environmental factors in controlling and handling bulls has already been discussed. There is a need for quantitative studies on the differences in behaviour of castrates and non-castrates kept under identical conditions. Work on groups of six bulls compared with castrates indicates that, similarly treated, groups of bulls show no increase in aggression (Wilkinson 1971).

From the production point of view there appears to be little doubt that bulls grow faster, show better conversion rates and an increase in muscle to fat ratios over the castrated animals (Champayne et al. 1969; Klosterman et al. 1954; Turton 1962; Baca et al. 1965). Wilkinson (1971) shows that on good grazing groups of young bulls show better production than castrates. On low nutritive diets, however, such effects may not be apparent (Eloff et al. 1965).

There appears therefore to be a real economic advantage as well as improvement in animal welfare (see methods of castration, below) in raising uncastrated cattle. Until nine months of age there is little likelihood of adverse behavioural effects. If animals are to be kept beyond this age, which may be a profitable and desirable practice, proper behavioural studies will be necessary.

Pigs. Walstra & Kroeske (1968) and Wismer-Pedersen (1968) reviewed the literature on the effects of castration on meat production in male pigs. They concluded that boars have a more favourable conversion rate than hogs and show a greater carcass length, smaller back fat thickness and less percentage of fat. The percentage of prime ham and shoulder cuts and dressing-percentage was higher.

However, boar meat after puberty has an unpleasant taste and odour. It is suggested that this is due to an accumulation of androsterone in the fat (Perry, personal comment). Walstra & Kroeske (1968) found that removal of spermatogenic tissue would also have the same effect. This second method is a very delicate operation. The substance responsible for boar taint has been isolated from the preputial glands (Dutt et al. 1959), although this may only be a reservoir (Perry, personal comment). The importance of these glands in courtship and successful breeding has recently been shown (Perry et al. 1972).

At present, therefore, if pigs are being raised for bacon (post-pubertal) the male must be castrated or implanted with DES (diethylstilbestrol) to reduce meat taint. On the other hand, there appears to be no reason to lose the extra carcass quality and growth rate by castrating pigs to be

slaughtered pre-pubertally (e.g. porker pigs). How large the differences
are between entires and castrates at porker weight has not been
ascertained. Entire porkers are unlikely to show marked behavioural
changes, except possibly an effect on feeding. It has been found that
boars restrict their own feeding to a greater extent than hogs, thus
increasing conversion rate and carcass quality (Walstra & Kroeske 1968).
At the very least, keeping porker pigs entire would be of benefit in that
they would not show the loss in weight gain and appetite which normally
occurs after castration in young pigs.

Horses. Horses and ponies are rarely kept as entire animals other than
specifically for breeding. Again the main reason given for this is that
stallions are more aggressive. Stallions kept isolated and used only as
'teasers' or for breeding do become very difficult to manage. However,
most mares or geldings kept under similar conditions (that is, isolated and
often overfed and under-exercised) behave similarly. Tractability of
stallions is shown by the Arabs who, because of religious laws against
castration, do not castrate their horses. These stallions are kept singly,
living with the family, or in groups, with no more trouble than mares
(personal observation). Care is taken to isolate oestral mares when
working with stallions. Dawson (1970) showed that handled well
thoroughbred stallions, notoriously reactive and difficult, could be ridden
in company with mares, alone or by children, or even driven, with no
behavioural problems. Such use of the stallions did not detract from the
animal's performance as a breeding male.

When working, stallions have the advantage of greater strength and
staying power than mares; they also look finer and carry themselves
better. This is used to good advantage by the famous Spanish Riding
School in Vienna who use only stallions for their 'haute école'. As riding
horses they are more predictable than mares since oestral mares frequently
behave unreliably. The author's experience is that both thoroughbred and
Arab riding and breeding stallions treated as ordinary horses and kept in
groups do not become vicious or more difficult to handle than mares or
geldings. Thus it is suggested that horses intended for certain types of
work such as jumping, dressage, racing and farm working should not
necessarily be castrated. Further and better use could be made of
breeding stallions for riding and other forms of work. At present such
animals are often used exclusively for breeding, which means that they
are not selected on the grounds of personal performance and behaviour
but on performance of relatives and in particular on conformation
criteria. Selection on the grounds of performance and trainability hand in

hand with conformation should lead to improvements and more tractable stallions.

Conversion rate, growth rate and even total intake have not been studied in the horse.

Sheep. Due to the small size of the ram, relative timidity and the fact that to date sheep have usually been kept at pasture and consequently handled little, there appear to be no behavioural advantages or disadvantages attributable to castration. However, work on the production effects is extensive. The picture that emerges on some effects is confused largely because methods of husbandry, which are important in determining the relative performance of castrates and entires, are not often given.

There is considerable evidence to indicate that castration has a detrimental effect on weight gain in sheep. Thus Clark (1965) showed that ram lambs were 1½ to 2lb heavier than wethers; Dun (1963) found rams to be heavier than wethers at 17 months and Wilcox (1968) showed a higher live weight gain in entires. Korotkov (1966) found larger weight gains and concluded that food conversion was more efficient in rams at pasture than wethers. In addition, Clark (1965) found that entire adults showed a more efficient pastoral use than lambs and wethers.

In relation to the effects of castration on wool growth the evidence is conflicting. Slen & Connell (1958) and Couthcolt (1962) found that males grew more wool than castrates. Von der Ahe (1966) also found that wool growth was greater in the entire animal, although the Baiburcjan method of castration (which is the squeezing and rupturing of the testes without destroying the epididymis) yielded an equivalent amount. However, the wool from these animals had an 'undesirable refinement'. On the other hand Dun (1963) showed that wethers produced 20% more clean wool than the rams, presumably because of more sebaceous glands in the ram. The ram wool was not only more greasy but more 'crimply'. Sahmardanov (1967) also found the wool quality of wethers better than that of rams.

It is well known that castration increases the amount of fat on the carcass of sheep (Clark 1965). It is also said to affect the ratio of bone to total weight. However, Tugai (1967) found that this was not so. Yeates (1965), summarizing the evidence, concludes that the ram carcass is of better quality in general than that of the wether. On the other hand Wilcox (1968) found ram carcasses to be slightly inferior in quality compared with carcasses of various castrates. Mocalovskii (1963) found the dressing percentage to be higher in partial castrates.

Thus in sheep there appears to be little production advantage in castrating the males. The main undesirable effect of not castrating appears to be that of having a large population of potentially breeding males (see B above). Where lambs go to slaughter before reaching puberty or the animals are housed this ceases to be an important objection.

Age and methods of castration
Complete removal of the testes is the traditional method of castration. There is considerable wastage and drop in weight gain of calves castrated by simple methods such as the elasticator, by comparison with open surgery of the 'burdizza clamp'. Fenton et al. (1969) showed that the elasticator method was often unseccessful or encouraged septicaemia. Growth rate and conversion rate depreciated during this period. Mullen (1964) showed this effect to extend for two months. Open surgery, although more successful, requires the services of a veterinarian. The burdizza clamp method of castration causes little inconvenience but requires skill in performing. An easy and reliable method of castration therefore does not seem to exist. The Russians invented a method of squeezing and rupturing the testes without destroying the epididymis which therefore allows continued circulation of testosterone (Baiburcjan method). This method is generally found to have better results in terms of performance than complete castration (Mocalovskii 1963; Creswell et al. 1964). Kosyh (1962) shows that such a method of castration results in comparable performance with the entire animals, and Dimitrov & Neicev (1964) and Mocalovskii (1963) show this to be true also of wool production. Von der Ahe (1966) mentions that partial castration of rams results in flock 'unrest', since there is regeneration of testicular tissue and the males once again become sexually active. Since partial castration appears not to show any improvement in performance by comparison with entire rams, the operation seems therefore to be of little advantage.

In sheep, late castration is reported to be better than early castration for performance (Campbell & Bosmans 1964; Chmielnick 1965), although these various castration groups were not compared with entire animals. In addition Yagav (1962) found late castrated calves and entires compared favourably in condition, conformation and weight with calves castrated at one month of age. An argument for late castration must, however, consider the added time, expense and skill needed for this as well as the likely increase in pain and trauma to the animal. In addition, anaesthesia is required by law in Britain for horses from birth, cattle and sheep after three months and pigs after two months.

Hormone therapy for castrates

Hormone treatments have been introduced to overcome the loss of weight gain in beef animals as a result of castration. Gassner et al. (1958), for example, found that when castrates were fed diethylstilbestrol (DES) the weight gain was increased by ½lb per day (a 12% increase). However, this does not completely compensate for castration, as entire animals were found to show an increase of 15 - 17% over untreated castrates (Wilkinson 1971). A disadvantage (particularly true of DES pellets) was a reduction in carcass quality. Thurber et al. (1966) treated early and later castrated range cattle with DES and found that the early castrates gained 21lb more than those castrated later. Unfortunately, bulls were not compared in their final assessment. Hexoestrol was used on partially castrated and completely castrated beef animals by Forbes et al. (1968). They found that hexoestrol had no effect on partial castrates, although it increased the percentage of fat in the carcass of the complete castrates. It must be considered here that untoward effects may be prevalent; Fletcher & Short (1974) found that oestradiol re-established copulation in castrated male red deer.

Androgen therapies might be more appropriate. Anabolic androgens with minimal effect on second sex characteristics such as androstendione are available and used on some cattle. The behavioural effects of these treatments have not been studied. Effects on aggression, general activity and sexual behaviour would be expected from analogy with other species.

Conclusions and recommendations

It can be recommended that further behavioural studies on castration in cattle, horses, pigs and sheep, *unconfounded by environmental effects,* should be carried out. These would compare groups of entires and castrates in order to assess what, if any, behavioural differences there are. Such work is particularly needed with cattle and sheep.

Further research work is also needed in stockmanship and the ability to handle and train animals.

Thus, with the existing knowledge on castration and its effects, it can be recommended that all males which it is not intended to keep beyond puberty should not be castrated, as performance is usually reduced by such a practice. In addition, animal welfare will be improved and economies in time and labour can be effected. This would apply to some porker pigs, fat lamb, veal and some barley beef. Post-pubertal housed beef and lamb should not be castrated, although boars may have a meat taint and therefore probably should be. This is provided butchers can be educated to overcome prejudices against entire male meat. There would be a further advantageous effect on the fertility of the female of having at least one entire or vasectomized male on the farm.

It is also suggested that quality horses which are intended for farm work, driving, the show ring, jumping, racing or dressage work should not necessarily be castrated, provided they are not handled by novices. Children's ponies, riding school horses and amenity horses should be castrated. More use of quality stallions other than for breeding would improve selection of appropriate behavioural traits and reduce the difficulty of handling isolated, under-experienced stallions. The maintenance of stallion licences with the inclusion of some assessment of temperament would ensure that stallions have a reasonable standard of health, conformation *and* temperament.

Where castration is recommended it should be carried out when the animal is as young as possible (and preferably by a veterinary surgeon or skilled technician) to avoid trauma to the animal and septicaemia thereafter.

Treatment of castrates with hormones, and partial castration, show neither production nor behavioural advantages over the entire animals. Such treatments therefore appear to be unnecessary although further work on the behavioural effects, particularly of hormone treatments, is needed.

CHAPTER 10
THE EFFECTS OF EARLY EXPERIENCE ON LATER BEHAVIOUR

This is a vast subject on which there is a considerable volume of experimental work, particularly in laboratory rodents and dogs and the domestic fowl. Little attention has been given to the effects of early experience on later behaviour and production in the larger agricultural animals with the exception of some recent work with cattle. All it is possible to do here is to mention briefly some works that indicate the scope of the effect of the early environment.

Sexual behaviour. The possibility of the attachment of sexual behaviour to members of another species has already been mentiond (cf. maternal behaviour). This is probably one of the most important effects of early experience which can cause problems in agriculture. Where animals are raised with siblings this behaviour is rare. However, when they are raised in isolation (e.g. calves in pens) it can be a problem. Imprinting on the husbandry man could perhaps become useful where AI is used at sexual maturity. If the heifer is imprinted on the stockman, it is possible that it will respond more appropriately to insemination with consequent increase in conception rate.

Fearful responses. These involve fleeing, avoidance, submission, immobility and shivering. It has already been mentioned that isolation appears to increase fear in young animals. In addition Levine (1962) showed that handling reduced the latency (time) to leave the start box on a raised runway; that is, reduced fear in rats. Early experience affects this behaviour in the group. Thus early loss of encounters will result in more 'submissive' (i.e. fearful) animals (see page 26). Donaldson (1967) showed that dominance and submissive tendencies of dairy calves could be related to early feeding and rearing conditions. Calves fed separately were more 'submissive' than those fed together.

Maternal behaviour. There is some evidence that the experience of having a mother affects the animal's ability to be a mother. This has already been discussed (page 58). Donaldson et al. (loc. cit.) rather surprisingly showed that cows that were raised separately as calves were better mothers (i.e. rejected young less and licked them post-parturiently more) than those raised in groups. The group-reared animals also vocalized more and failed to allow nursing. Dairy cows' maternal behaviour has to some extent been selected against and individuals do vary greatly (Kiley, in preparation). Thus such a result needs replication before its relevance to husbandry can be confirmed.

Aggression. Levine (1957) showed that handling early in life decreased the latency of attack on another mouse. Again, this is a subject on which we need more information.

Reactivity. Some animals react more quickly to environmental change and they are more easily excited. It appears that handling early in life can reduce such responses (termed 'emotional responses' e.g. Denenberg 1964; Levine 1962; Ader & Conklin 1963). The absence of handling increases these emotional responses (e.g. Thompson & Schaffer 1961). These authors do not always mean similar things by handling, nevertheless, anecdotal evidence with dogs, horses and cattle confirms the suggestion that early gentle handling and caressing will affect reactivity and trainability. Another factor here is that such handling by man increases early socialisation to him, which may well itself act as a foundation for future relationships between man and the animal.

Other types of environment experienced early in life will probably also be shown to affect general reactivity and emotionality (e.g. isolation, crowding, etc.).

Effect of early environment on feeding. Marx (1952) showed an increased rate of eating in the adult rat as a result of infantile food deprivation and Elliot & King (1960) showed a similar result in puppies, although the older faster-eating puppies did not eat more than the earlier non-deprived animals.

At this stage there is very little experimental work on the effects of early experience on later behaviour in farm animals. Nevertheless, by analogy with other species, particularly primates (e.g. Harlow et al. 1963), it is more than likely that the physical and psychological environment of the young farm animal is very important for the development of adaptive behaviour and hence particularly careful consideration must be given to its design—but too little is known.

swine to be olfaction, audition and vision. From pilot work using auditory conditioning to different tones in pigs for food reinforcement, Kiley (unpublished data) found them to have learnt to discriminate between 2 notes, 4 tones apart, in 4 to 8 trials, which is a fast rate of learning. In addition, vision was used a great deal by these experimental animals. Kudryartzev (1962) showed that cattle also used visual, auditory and olfactory cues in classical conditioning experiments.

D. Use of positive or negative reinforcement

Circus trainers and most animal handlers and trainers use largely positive reinforcement; only rarely, and after prolonged training with positive reinforcement, do they resort to punishment. Such techniques could certainly be adopted for the best results with animal conditioning in the farm situation. From the practical point of view, it is also much easier to arrange for the animal to work for food, than to shock it or use some other form of negative reinforcement.

CHAPTER 12
STEREOTYPIES

A stereotyped behaviour is usually defined as behaviour which is repeated
in exact detail and is purposeless (Fox 1965) The use of the word
purposeless is inaccurate because it assumes greater knowledge concerning
causation and motivation than we have at present. To qualify it by
'apparently' purposeless is misleading as some stereotypies may be
exaggerated forms of clearly purposeful behaviour (e.g. stereotypic feeding
in chicks which is repeated pecking at food pellets). Thus we are forced
back to a variant of the dictionary definition which defines stereotype as
"an aberrant behaviour, repeated with monotonous regularity and fixed in
all details".

Although stereotypies are not particularly common in commercial farm
animals today, they are not unknown. In some cases they interfere
considerably with production (e.g. inter-suckling of calves and lambs, crib
biting in cattle and horses, chewing and rubbing of pigs and calves, wind
sucking and weaving in horses, fleece biting in sheep). More important is
the consideration that the occurrence of stereotypies can act as symptoms
of inadequate environmental design, which may be affecting production
through physiological pathways.

For various reasons (e.g. mutual interference, ration feeding and
economy), animals are becoming further confined and restricted in free
movement (e.g. sow stalls, barley beef units, veal calf crates and battery
hens). One of the behavioural results of such practices may well be an
increase in the development of stereotypies, as well as other behavioural
changes discussed elsewhere (e.g. increase in aggression). Little
experimental work has been attempted but it is useful at this stage to
review the literature to date, much of which is on zoo animals.

Some species, or individual animals, may be more liable to stereotypies
than others (for example, canids, felids, primates and non-ruminant
ungulates, such as horses, elephants and rhinos which more often develop
them, than ruminants such as antelopes, tylopods, cervids and bovids).
However, any animal is lidely to develop a stereotype if exposed to the
various appropriate conditions. These are:

(i) Restricted movement and insufficient floor space. Levy (1944) found
that the amount of stereotyped head shaking in battery hens was directly
related to the amount of floor space. He then quotes many observations
of 'tics' and stereotypies performed as a result of movement restraint of

humans and animals in zoos. Keiper (1970), however, describes two stereotypies of caged canaries: (a) 'route tracing' (repeatedly following an imaginary route with the head and body) which he showed can be reduced by increasing the size of the cage, and (b) 'spot pecking' (pecking at spots in the cage repeatedly), that was not.

In the stalled horse, weaving, crib biting and wind sucking seem to be related to movement restriction and can be reduced by putting animals out to grass (personal observation). The sow in sow stalls may develop rubbing of the nose or back against the sides of the pen which may become a stereotype with pathological implications, as can stereotypic licking in confined dogs (Fox 1965a). The pacing up and down the bars of many caged felids, canids and ruminants also appears to be related to restricted size of cage (Meyer-Holzapfel 1968).

(ii) Sterile, dull, constant environment. The development of a stereotype appears to be particularly prevalent in sterile or dull environments. Thus Levy (1944) showed that hospitalized children deprived of toys performed pronounced stereotypies. These were stopped when the toys were returned to the children. Hutt & Hutt (1965) showed that stereotypies were reduced in autistic children when toys or adults were present in the environment. Berkson et al. (1963) showed that the opportunity to manipulate the environment was more important than the size of the cage in caged primates.

Fox (1968a) showed that frequent changing of the cage of his dogs reduced the performance of stereotypies. Nissen (1956) stated that boredom controlled the performance of stereotypies in his chimps. Clearly, as pointed out by Welsh (1964), there does appear to be a "mean level of environmental stimulation". If there is either too much or too little sensory input, then behavioural changes are likely to result. One of these, particularly as a result of insufficient environmental stimulation (boredom), is the the performance of stereotypies, which may lead to self-generated increased sensory input. Too much or too little environmental stimulus, therefore, might be expected to encourage the development of stereotypies.

Meyer-Holzapfel (1968), working with bears, found that after acquisition of a stereotype increasing the environmental stimuli (by cage washing or someone entering the cage) resulted in an increase in the rate of performance performance of the stereotype.

(iii) Novel environment. Berkson & Davenport (1962) and Berkson et al. (1963), working with mental defectives and chimpanzees respectively, showed that the introduction of a novel object into the environment

reduced the performance of stereotypies. However, Keiper (1970), working with canaries, pointed out that this effect lasts as long as the bird is not habituated to the novel object, that is, until it loses its novelty. He introduced unknown canaries into the cage of the canaries that performed stereotypies and found that the stereotypies were reduced until the animals became habituated to each other.

(iv) Frustration and conflict situations. There is some evidence to suggest that established stereotypies are performed when the animal is in a frustrating or conflict type of situation. Thus Meyer-Holzapfel (loc. cit.) found that a rhino isolated from its mate in an adjacent cage performed stereotyped behaviour. It has often been observed that stereotyped activity by zoo animals is more common before feeding time or when the animals see and smell the food but cannot obtain it: a food frustration situation. Kiley (1969) shows experimentally that stereotypies in the horse could be elicited in a food frustration situation, and Duncan & Wood-Gush (1972 and 1974) induced stereotypic pacing in the domestic fowl as a result of food frustration. These authors argue that, since this pacing was reduced by the use of a Rauwolfia tranquillizer which also reduced typical fear responses, these movements were therefore probably motivated by fear. In rats and dogs, however, stimulants usually induce or increase the performance of stereotypies, whereas sedatives reduce them (e.g. Schiørring & Randrup 1968; Willner et al. 1970). However, these drugs also have effects on a very wide range of other behaviours and it is therefore rash to consider that such actions are primarily motivated by fear. The effects of drugs on the performance of stereotypies more obviously relates to 'general arousal', which is increased by stimulants. One of the results is an increase in the performance of stereotypies, while tranquillizers decrease them.

(v) Absence of key stimuli. The most common example of a stereotype apparently caused in this way is the development of inter-suckling in bucket-reared calves and lambs (e.g. Stephens & Baldwin 1970). Here there is an adoption of a substitute for a key stimulus. Inter-suckling, usually in the umbilical or scrotal region in bucket group-reared calves, is a problem of very considerable economic consequences. It is usually the suckling calf which suffers most from reduction in intake or swallowing of urine. However, infection in the umbilical or scrotal region or even the ears may well follow in the suckled calf. All of these effects at best cause loss of condition and growth rate and at worst death.

Wood et al. (1967) did a survey of inter-suckling in dairy herds in England and Wales and confirmed that inter-suckling is imitative. He found that, when present, it occurred in at least 27% of the calves. It can continue into adulthood. Guernseys are said to be particularly prone to develop this habit, although Wood describes a herd of Friesians that had to be disbanded because it had ceased to be economic as most of the cows were drinking their own or other's milk!

Wood found the most effective remedy to this problem was the individual isolation of the offenders up to 9 weeks old. However, it would be unwise to advocate isolation rearing of calves since Coffey (1971) has shown that isolation reduces growth rate. Wood showed that dry feeding was also effective in curing this problem. This again may be no solution as intake may be depressed by such a rearing technique (see page 40). Repellents on the suckled calves were less effective. On the other hand, the earlier the calves were weaned the less they tended to inter-suck. Others maintain that tying the calves before and after feeding reduces inter-suckling (personal comment, Farm Buildings Investigation Unit, Aberdeen).

This is a problem where more research is urgently needed. It is possible that group size, type of housing and bedding may have some effect on its development, as well as the more obvious factor of the type and availability of feed.

One new approach would be to try and replace the key stimulus in an adequate enough form so the calves or lambs do not suck on each other. Teats placed on the wall of the pen have been tried in this context, not always with much success, although artificially reared lambs will suckle their empty feeding teats for a considerable proportion of the time. The effect of having, for example, a warm teat with a hair surround, and the angle at which it is presented on initiating suckling in the calves should be investigated. In this way adequate sucking substitutes could be designed.

Pen chewing may also be described as an activity occurring in the absence of key stimuli. Isolated calves may suck or bite bits off the pen and this may continue into adulthood (Ewbank 1962). Pen chewing is common in barley beef and veal units where no fibre is given to the animals. In order to prevent this, and to allow for normal rumen function and rumination, various plastics and paper supplements have been tried in the diet.

These are the types of situations which are at present known to induce stereotypies. It must be borne in mind that normally several of these factors are present at one time. The stereotype, therefore, may well be the result of several summated factors.

(vi) Stereotypies and conditioning. Stereotypic ear scratching can be

induced in the rat by placing grease in the ear. The stereotypic response continues well after all external stimulation has ceased. Lal Harbans & Robinson (1965) developed stereotypic bar pressing in the rat through operant conditioning which then became generalised and occurred in many different situations. In addition, traumatic environmental change apparently reduces the performance of stereotypies. Friedberger & Frohner (1904) quote Panecchi, a cavalry officer, who said that after a particularly traumatic battle in 1866 stereotypic crib biting and wind sucking, habitual in his cavalry horses, was suppressed. This is reminiscent of Pavlov's dogs which, after the traumatic Moscow floods, no longer showed their conditioned reflexes.

This learning hypothesis for the development of stereotypies shows many parallels with the development of displays (Kiley 1969, page 346). An activity such as head tossing in the pig and horse is initially evoked by some mild stimulus (such as nasal irritation as proposed by Cook 1971 in the horse). It becomes accidentally reinforced and thereafter becomes associated with that particular situation, such as head throwing during exercise in race horses. Subsequently it may generalise to be performed in many different situations and show an emancipation from its original causation.

Berkson & Mason (1964) arrived at an explanation for the performance of stereotypies in terms of 'arousal', since neither novelty nor spatial and visual restriction alone controlled the amount of time monkeys spent doing stereotypic acts. Because of undemonstrated physiological implications of this term, 'excitement' might be more satisfactory.

If the cause of stereotypic behaviour remains complex and somewhat obscure, its function may be more apparent. It could be that a very low level of sensory input is as unsupportable as a very high level. Therefore, in either situation—one of very low sensory input when the animal is bored or one of very high sensory input when the animal is excited (for example, an environmental change, introduction of strangers, conflict or frustration situation)—stereotypies function as an adaptive mechanism to switch attention from the unacceptable environment to self-stimulation.

Finally, it must be pointed out that an established stereotype may well occur in a greater variety of situations from that which would allow its development, as some animals with very well established stereotypies will perform their action (in many different situations) almost without interruption. Duncan & Wood-Gush (1974) show how difficult they are to overcome once they are established. Some of the contradictions in the literature may be the result of not drawing a firm distinction between developing and established stereotypies. For example, Hutt & Hutt (1965) found that increasing the environmental complexity for autistic children increased the performance of established stereotypies, whereas this has been shown to decrease the performance in children where the stereotypies are not so well established (Levy 1944).

For practical application it must, therefore, be clearly appreciated that to change the environmental conditions to cure an established stereotype may well have little effect. However, the provision of an 'adequate' environment of sufficient complexity to avoid insufficient sensory input may prevent its development.

Thus it is suggested that the performance of stereotypies should be regarded as a symptom of environmental inadequacy.

CHAPTER 13
FOOD INTAKE AND ITS CONTROL

This is a field in which much research effort has been directed as production is directly affected by food intake. Again, however, the behavioural aspects of intake have largely been neglected. The optimal amounts of starch, protein and minerals to feed for a particular end product are known. The problem is, however, to ensure that each individual eats these amounts. The environment and social controls of food intake must be considered in more detail in order to answer this.

Palatability and taste. Palatability appears to be a general term which depends in particular on the taste, the nutritional value and the physical form of the food. Unfortunately, most of the work to date has confounded these three independent variables.

It was suggested early on in the history of agricultural research that animals show 'nutritional wisdom', that is, given a free choice they would select a balanced diet and the minerals which they require. There is clearly some truth in this approach, since animals low in certain minerals or requiring large amounts of them will go to considerable length to fulfil their requirements. It has been demonstrated that palatability has a very important influence on selection of dietary requirements. For example, Scott (1946), working with rats, found that some rats failed to eat diets containing casein, preferring the less nutritional diets of hydrogenated fat, sucrose and salt mixtures. Further work (Scott & Quint 1946) showed, however, that offered a choice of only different proteins few rats refused all choices but they did show preferences for some (e.g. casein, lactalbumen and fibrin) over others (e.g. egg albumen). They conclude that this indicated that there was no specific 'appetite for protein' in the rat but rather simple preferences of palatability. Gordon & Tribe (1951) found that when pregnant ewes were offered the choice ad libitum of diets consisting of different proteins or carbohydrate concentrates, hay, minerals and water, they failed to select a ration which allowed them to bear and rear strong, healthy lambs. Similarly, Kare & Scott (1962) found that chickens showed distinct preferences for food which were not related to nutritional value. For example, they greatly preferred 'corn' (that is, wheat) to barley, although they were similar in nutritional value. However, various workers have shown specific appetites for induced deficiencies in the domestic fowl. For example, Hughes & Wood-Gush

(1971) showed a specific appetite for thiamine, and Hughes & Dewar (1971) for zinc. Wood-Gush & Kare (1966) induced an appetite for calcium in deficient birds and Hughes & Wood-Gush (1971) found this to be a learnt preference and not an unlearnt homeostatic control. An appetite for sodium was not induced by deprivation (Hughes & Wood-Gush 1971).

The masking of unpalatable foodstuffs in order to induce acceptance is used frequently (particularly with sugar derivatives) from children's cough mixture to animal foodstuffs (e.g. urea with molasses, high cellulose grass or hay with molasses, Kare et al. 1965; Baldwin 1969). The presentation of a 'pleasant tasting' foodstuff may result in it being taken in the place of a nutritionally balanced one and in some cases such preference continues until death from a nutritional deficiency (Tribe 1950b).

However, before we can develop such techniques further, we must have a more complete understanding of taste preferences unconfounded by physical characteristics of the foodstuff and so on. Also a comprehension of individual preferences in foodstuffs (e.g. Blaxter & French 1944 and Kare et al. 1965) is needed.

Temperature can affect palatability. Kudryartzev (1962) showed that the taste reactions of dairy cows were reduced below $5^{\circ}C$. Thus the feeding of cooled food might allow substances to be eaten that would not normally be. What the effect of temperature of foodstuffs on total intake would be, however, still has to be investigated.

An understanding of 'palatability' and taste preference is becoming more and more important in the manufacture of animal foodstuffs as substances such as yeast grown on petroleum by-products come on the market to be used as animal foodstuffs. Leaf protein, soya bean meal, urea, newsprint and chicken manure are some of the unconventional foods that are becoming more important in the manufacture of animal foodstuffs.

Physical form of the food. The physical form of the food will affect palatability and intake. For example, pigs fed meal will eat less than those fed pellets; similarly, calves reared on *Cynodon dactylon* (an indigenous African grass) in different forms ate more of the pellet form by 32% and were faster growing than those fed the long grass. However, when concentrates were mixed in the pellets, the rate of gain and consumption was less than in the animals fed the ground or long form. The addition of 1.5lb of straw increased the consumption of the pellets by 26% and the rate of gain by 37% (Cullison 1961). Thus it appears that the feeding of meal only may reduce intake. However, if all foods are offered pelleted, intake may also be reduced. Behavioural problems related to the reduction of rumination will also affect ruminants on such rations.

Campling & Freer (1965) showed similarly that intake of some foodstuffs is increased when the size of the particles is decreased. Thus grinding straw will increase intake 26%; the intake of dried grass on the other hand is not affected by whether it is ground or not. Urea, although a cheap protein food, is not readily taken by most agricultural stock. However, when urea was disguised in pellets, Campling & Freer (loc. cit.) showed that the intake could be increased by 53%.

The texture of the food can have effects on other behaviour; for example, it may have an effect on cannibalism in hens. Bearse et al. (1949) and Skoglund & Palmer (1961) found that this was more common when the birds were fed pellets than mash. This was not found by Ziegenhagen et al. (1947) however.

Digestibility. One of the major philosophies that control the work on food intake and selection in agriculture today is the idea that total intake and selection is related to digestibility (Blaxter 1967). This is not disputed here; nevertheless, it is the opinion of the writer that this view has often obscured the importance of other environmental factors that control intake. The importance of the effect of digestibility of food in controlling animal movements in wild as well as agricultural animals has in particular been over-emphasised.

An example of the fact that digestibility is not the most important quality controlling food intake is given by Porter (1953). Dairy cows were presented with three different types of roughage: grass silage, corn silage and hay. Although more corn silage was often eaten, the animals presented with a choice took advantage of this choice. They did not stick to the most digestible type. Unfortunately, Porter does not mention whether the total intake was increased where the animals were given a choice of foodstuffs. He does, however, mention that other things such as weather, shelter and available water affected the choice of roughage.

The weather. Weather can control food intake to a very considerable degree in free-range sheep. For example, intake will be severely restricted in sheep in bad weather as they will not move far from shelter (Kiley 1974). The grass around the shelter will be overgrazed and thus intake restricted, although an infinite amount of grazing may be available elsewhere.

Temperature. Lowered temperature is said to increase food intake in most farm livestock and, more important still, it is believed to reduce conversion rate. For this reason apparently, it has become accepted practice to try and control the environmental temperature as far as possible. Any such advantage often bears little relationship to the capital expenditure necessary for such a practice. Thus Noland et al. (1965) showed that although there were the fastest gains in pigs fed ad libitum in an enclosed system these differences in weight gain and feed conversion were not significantly increased over the open system. If extra costs of the buildings were to be taken into consideration, business men would wonder why such housing has become so common!

Monotony. Balanced manufactured rations are normally fed to housed animals today. Thus there is little variation in taste or consistency of the pelleted or milled ration. It is possible that the monotony of the diet restricts intake. This has not been studied quantitatively. It is evident that the monotony factor may also have some effect on grazing. Cattle, after some time on the pasture, consistently chose to graze the hedgerows and hedges rather than the homogenous grass ley (Stapleton 1948 and Dr Greenhalgh, personal observations).

The possibility that monotony may restrict intake is indirectly recognised by offering a very varied concentrate ration to animals where high intakes and growth must be encouraged. These rations often consist of flaked maize, barley, cotton-seed cake, oats and bran to, for example, calves and foals, and a similarly varied diet to puppies.

Frequency of feeding. Ungulates in a wild state normally do not obtain all their food in one feed. In current agricultural practice, however, once a day feeding is becoming more economically attractive. It might be expected that such a practice would reduce growth rate and production. There is little work on this but what there is with ruminants (e.g. Faichney 1968 in sheep) suggests that frequent or twice daily feeding does not affect benefits substantially. Faichney (loc. cit.), however, showed an increase in body water (and, therefore, body weight gain) with 3 hourly feeding of sheep as compared with 24 hour feeding.

Once a day feeding might be more likely to affect intake in non-ruminant ungulates, however, where only one stomach is available to be filled. A golden rule of horse management is frequent feeding but the effects of this as opposed to once a day feeding have not been scientifically assessed.

Chickens under intensive management normally have access to food whenever they like. Duncan et al. (1970) found that each bird had its own feeding rhythm which was divided into bouts or meals but there was no characteristic size of meal or length of interval, although the larger the meal, the longer the interval before the next.

Most other livestock under intensive management are food restricted and their diurnal rhythms tend to be very controlled by these feeds (M. Kiley unpublished data on beef cattle and veal calves). In veal calves, excessive calling and other signs of hunger and distress are evident during the period of adaptation to the unit. The effects of restriction of roughage in conjunction with restricted feeding, thereby causing the animal to have a great deal of 'free time', might lead to various pathologies (e.g. stereotypies) and/or reduced production. This is a subject that needs further work.

Food dispersion. Given adequate supplies cattle, for example, can be allowed 9 ins - 1 ft eating space/individual. This allows all to obtain their ration. By contrast horses must be allowed several yards to ensure each animal obtains some food (personal observation). Ewbank & Bryant (1969) found that some pigs would stand in front of feed trough preventing others obtaining a ration.

Good inter-species comparisons are not available to date but it is evident that homogeneity in size, sex and age of animals in a group is likely to contribute to smaller feeding area per individual, although on other grounds this may not always be the most successful way of keeping animals (see page 28).

The effect of drugs on intake. Many drugs decrease intake (e.g. neuroleplics). However, Bainbridge (1970) found that some minor tranquillizers (e.g. chlordiazepoxide, phenobarbitone and meprobanate) increased 'appetite' in isolated rats. This did not occur in the grouped animals, however.

One approach to behavioural problems on the farm is to advocate the wider use of drugs, particularly tranquillizers. Caution must be exercised in such an approach, however, since effects on other behaviours such as intake are not fully understood.

The effect of sex on intake. Recent work with pigs (Houseman 1973) showed that boars had a 7% lower intake than gilts and 12% lower than hogs, whereas the specific gravity of the carcass was 15% higher than that of gilts and 25% higher than that of castrates.

The relative conversion and production merits of entire versus castrate males has been reviewed elsewhere (page 60 et seq.). However, that this reflects on intake in an indoor environment is interesting. It is well known that intake decreases in entire males during the rut as a result largely of reproductive activity (e.g. Grubb 1974). Is this effect the result of different metabolisms, or that the males are more involved behaviourally in doing other things which decrease eating time? Or is it the result of using different methods of feeding? In chickens, testosterone treated chicks have been shown to search for food in a different way from normal males (Andrew & Rogers 1972).

Such questions, and a host of others relating to specific reasons for food preferences as opposed to blanket terms such as 'palatability' are awaiting answers.

CHAPTER 14

A REVIEW OF GRAZING BEHAVIOUR AND SUGGESTIONS FOR
USING SUCH KNOWLEDGE FOR INCREASING PRODUCTION[1]

Although James Anderson in 1797 pointed out that the behaviour of
the grazing animal was vitally important in the utilisation of pasture,
this fact seems to have been largely ignored since. There is an enormous
volume of work on the utilisation of pasture, which has been concerned
particularly with the nutritional quality and relative digestibility of the
various floral species. Only comparatively recently has it been shown
that an accurate estimate of the pasture eaten cannot be obtained by
cutting samples of the pasture, since the animals select particular plants
and stage of growth of plants (Weir & Torell 1959). Other behaviour
that affects grazing, such as defaecation, individual variation within group
and species, social facilitation of grazing and behaviour related to the
weather have been mentioned in the literature, but there is little
quantitative work to date. Clearly if maximum utilisation of a pasture is
required all the types of behaviour that interfere or affect grazing in one
way or another must be understood and carefully considered.

Grazing is essentially a behavioural problem, an understanding of which
would lead to more successful management of the grazing animals and
their various dependents, as well as of the grassland. Thus it is a
behavioural problem that interferes with management not only in an
intensive farm or wildlife park but also on a more extensive scale.
There are several excellent reviews on grazing behaviour written by
agriculturalists (e.g. Johnstone-Wallace & Kennedy 1944; Tribe 1950a;
Hancock 1953; Arnold 1962); however, to justify a further review let
me say that these works are somewhat out of date now and secondly,
as far as I know, there has not been one attempted by an ethologist.
The most important behavioural factors controlling grazing have been
listed and reviewed elsewhere (Kiley 1974, to whom the reader is
referred). An outline of them is given here.

The first problem is selectivity, which is far from understood. Animals
select grazing on the basis of plant species, part of plant, growth stage
and length of the plant. It has been suggested that selection is primarily
on the basis of digestibility (Blaxter 1962) but this is by no means a
complete picture. Palatability is a general term which is not understood.
Selection can also be on grounds of taste or physical form of the plant
and can be affected by nutritional demands, learning and other factors.

One of the most important factors affecting selectivity is, however, availability. Differences in age, sex and species of the animal ensure difference in selection.

Social facilitation of grazing has been reported anecdotally but the total extent of this is unknown.

The time spent grazing depends on availability and circadian rhythms. The weather and topography can also affect grazing. Social organisation and herding can interfere and change grazing patterns.

The placing of defaecations and subsequent avoidance of them when grazing is different for each species and has a very great effect on subsequent availability of grass and parasitic infection of the grazing animals.

A consideration of all these effects suggests that the management of grassland, either on an intensive or extensive scale, profits from multi-species grazing. Ways in which this could be done are outlined.

Selectivity in grazing. This has been known at least since Linneaus (1749) did a series of preference tests of different species; we do not know a great deal more in 1977.

As Stapleton (1948) pointed out, the degree of grazing selection is to some extent related to availability. Thus the more food available the more selective the animal will be. Another group of workers headed by Blaxter are convinced that digestibility is primarily important in selection (Blaxter 1962). There is no doubt that availability and digestibility are of great importance in selection by the grazing animal but they are by no means the only considerations; other behavioural factors of sometimes equal importance are very often overlooked.

That selection in grazing was a real problem in cattle was perhaps finally accepted following the publication of a paper by Hardison et al. in 1954. They showed that there was a greater selection by the *grazing* cow for crude protein, either extract or ash, than by the stalled animal fed on clippings from the same area. Weir & Torell (1959), using oesophagal-fistulated sheep, showed a similar result. This should serve as a warning to wildlife ecologists who tend to do preference tests with penned animals fed on clippings.

On what grounds then are the animals making their selection? 'Nutritional wisdom' is often assumed and movement of groups of animals to different areas is often explained on this basis. However, taste, physical form of the plant, learning and other factors confound any innate preferences that may exist; and they may or may not be related to the nutrient value of the grazed material.

Although we are far from understanding the complete mechanism controlling such behaviour rays of light are falling on it, mainly from detailed work with the rat. Thus all we can do at this stage is to indicate the variables which do affect selection in a grazing situation. Of course, different species and even different individuals make their selection on varying grounds. Selection on grounds of taste was demonstrated by Roe & Mattershead (1962) who made an extract of a palatable strain of *Phalaris* and sprayed it on an unpalatable one. The sheep thereafter preferred the sprayed strain. Selection on grounds of stage of growth has been demonstrated; for example, Bell (1970) found that Serengeti herbivores selected on stage of growth and locality. Milton (1956) similarly found that sheep in the Welsh hills would take species normally unpalatable when they had some succulent growth.

Specific parts of the plant may be selected. For example, Johnstone-Wallace (1937) found that cattle preferred leaves to stems. The length of the plant is also important as Johnstone-Wallace & Kennedy (1944) pointed out. Cattle prefer the grass to be 4 - 5 inches high whereas sheep and horses prefer it shorter.

Techniques for making these assessments have remained relatively crude and it is more likely to be nearer the truth to say that all these factors are important in selection by the grazing animals, the emphasis changing amongst species.

In addition, plant selection is to some extent dependent on the age of the animal. Leaver (1970) showed that heifers selected the tops of the grass, while the pregnant cows grazed the whole plant. Lambs and foals also tend to select the tops and seed heads, partly no doubt due to ineffiency of grazing.

The sex of the animal may equally have an effect on selection. It has been shown that testosterone affects the method of feeding in chicks (Rogers 1971). Testosterone treated animals search more persistently for a particular cue on which they have been trained; perhaps male grazing animals do the same. Certainly there is some evidence to suggest sexual differences in grazing behaviour of sheep (Foot & Doney 1971, personal comment).

Selectivity is to some extent conditioned by the variety and type of floral species available. Thus the catholic tastes of sheep shown by Linnaeus were confirmed by Doran (1943) on the high aspen ranges of America. However, Meyer et al. (1957) found that on ley pasture, sheep are more selective than cattle. In addition, Milton (1956) found that the mere presence of a disliked species can strongly influence the utilisation of a liked species.

Individual variations in time spent grazing was shown by Hancock (1954) with pairs of identical twins compared with the herd, which suggests that selection and amount grazed may also show considerable individual variation as production does (e.g. milk yield).

Learning is also important in grazing. Dove (1935) suggests that young horses, swine, goats and rabbits learn what to select from their mothers and Tribe (1950c), reinforcing such a view, showed that mortality in sheep was increased when they had no opportunity to learn selection from their mothers.

Social facilitation. Social facilitation of intake has, of course, been demonstrated in the rat. There is considerable anecdotal evidence of social facilitation of grazing, lying and possibly ruminating in herds of ungulates (Hardison et al. 1954; Dove 1935; Tribe 1950c; personal observation), but the extent of this has not been measured, even though this would be a very useful thing to know. Bouts of these activities tend to occur in the grazing mammals, but to some extent this may be confounded by circadian rhythms.

Time spent grazing and circadian rhythms. Hardison et al. (1956) showed that the time spent grazing was inversely related to availability of herbage. Time spent grazing may also be related to total dry matter intake (Castle et al. 1950), cellulose content (Lancashire & Keogh 1966) and such factors as the weather. However, there is a maximum limit to the time that will be spent grazing even when herbage is scarce. From the management point of view, particularly of marginal or overgrazed land, it would be useful to know where that maximum is for each species. It would then be possible to prevent loss of weight of animals while ensuring maximum usage of the pasture.

More important for the management of intensive systems is that there is very likely a minimum grazing time. Thus the animal may be able to consume all it requires for both maintenance and growth in a very short time on a good ley pasture, but it will continue to graze, thus wasting food, in the same way as happens with concentrate feeding.

Time grazing may be a very rough estimate of herbage availability at the extreme ends of the scale, but it is too gross and inaccurate a measure where availability is adequate. Intake and time grazing will be influenced in addition by the efficiency of grazing, for example, the number of bites per minute, size of mouthful and so on.

Earlier than the 1950s it was assumed that cattle did not graze at night. Thereafter Castle et al. (1950) and others, as a result of observing animals through the full 24 hours, found that they grazed almost as much at night as they did during the day (60% during the day, 40% at night). This discovery profoundly affected management as thereafter cattle were placed on good grazing at night as well as during the day. With more detailed investigation it became apparent that there were circadian rhythms in grazing. Thus cattle, for example, tend to graze in the evenings and often late at night, early morning and possibly mid-morning.

Monotony factor. It has been observed by several workers (e.g. Stapleton 1948; personal observation) that cattle, sheep and horses, at least, frequently graze hedgerows and rough pastures some of the time in preference to rich leys. It is possible that leys consisting of one, two or three species heavily treated with fertilizer or slurry become monotonous in taste and a change is sought, even if the chosen herbage is less succulent and less easily digestible. The effect of this possible monotony factor on intake is another interesting topic for research.

The weather and topography. Hunter (1964) points out that the amount of grazing and resting time in sheep is largely controlled by the weather, as are the movement of groups of hill sheep. Hardison et al. (1954) showed that time spent grazing, lying, ruminating and loafing depends to some extent on the temperature and location in cattle. However, the amount cattle graze is generally considered to be less controlled by the weather than that of sheep but this has not been recorded, to my knowledge, under equally inclement conditions.

Sheep will not eat if this involves going out of sheltered zones to graze in bad weather (Foot & Doney, personal comment). The steepness of the slope is also of particular importance as to whether the sheep graze or not. Thus sheltered, less-steep slopes are often overgrazed. This could well be a considerable problem in conservation areas where weather conditions are extreme.

The availability of water in arid areas frequently determines the grazing patterns of cattle, sheep and goats. The area around the waterhole tends to be grossly overgrazed with consequent erosion whereas this happens to a less extent where indigenous ungulates are grazed, as many of them are capable of going for prolonged periods without water (Lamprey 1963).

Defaecation. This factor must be considered here as soiling by faeces strongly influences the selection of grazing areas. Although it is becoming a very considerable problem in the intensively managed grazing areas, very little work has been done on the individual defaecation habits of any species. It has been estimated by Johnstone-Wallace & Kennedy (1944) that one cow produces 46lb of faeces covering 8 square feet during 24 hours. Cattle, horses and to a lesser extent sheep will not graze over their own defaecations hence, in a large herd of dairy cows for example, a large amount of grass will become essentially unavailable as a result of faecal contamination. With successive grazings during the same season the amount of grass of a ley actually utilised will decline. Proctor et al. (1950) measured this and found that 81% utilisation declined to only 43% on the third grazing of the season. Greenhalgh (personal comment) found that the area affected by the faeces in terms of rejection of the grass by cattle around the faeces was not due to any change in chemical composition of this grass, but rather to the continued presence of the faeces. Oldberg (personal comment) found the same for horses.

Various animals will, however, graze over other species' defaecations so that a rotation of animals (e.g. cows followed by sheep and horses) would reduce the wastage factor. If we knew more about the different species' defaecation habits and how this affects selection, more efficient management would result.

Parasites. The faeces are, of course, the main source of infection or reinfection by intestinal parasites in grazing animals, particularly young ones. Thus the avoidance by individual species of their own faeces has a beneficial effect on them particularly because, as Rose (1963) and others showed, many of these parasites migrate only a very short distance from the faeces. Again, we want more information on the behaviour of the parasites before we can work out really effective grazing schemes.

Social organization. Social organization interferes with grazing considerably and this can be a particularly important problem for conservationists. A good example of this is to be found in Hunter & Milner's work (1963) with South Cheviot sheep. These animals stayed in their own home areas (as do most hill sheep), although some of the animals were in areas which supported more palatable sward than the rest. Sheep from other groups did not migrate to those areas. Even when the animals were shepherded into folds and were released after several days, they returned to their own home areas.

Home areas, territorial areas, formation of harems and associations between individuals are bound to affect the free movement of individuals to graze. It is clear that, particularly in close-folded animals, dominance relationships may well affect the ability of certain animals to obtain the supplementary feed or the new grass at the electric fence or to approach the water trough.

The net effect of such behaviours, particularly on an extensive scale, is that to increase or decrease stocking rate will not necessarily lead to improved production on an area, although from the floral point of view it should. This is because of the non-random spread of the grazing animals in the area, which makes some sectors virtually psychologically unavailable to some individuals.

Herding. The herding of animals clearly affects grazing; a classic example of this is to be seen in dairy cows which are herded around two or three times per day. Even on an extensive scale on the range the intake of sheep can be substantially influenced by shepherding. Doran (1943) found that it reduced selection and even caused an increased intake of poisonous plants. Herding also decreased eating time and increased resting time as compared with non-shepherded animals on a similar area.

If these behavioural factors of grazing are considered, then certain types of management clearly suggest themselves. It is apparent that grazing is essentially a behavioural phenomenon, and to develop management techniques without considering most of the aspects of grazing that have been mentioned will probably result in unsuccessful policies. Of course, the controls are not so rigid on an extensive scale but, as the land becomes further enclosed, management becomes more essential. Many of these problems that agriculturalists are struggling with become relevant suddenly to the conservationist and safari park manager. If we take into account these behavioural factors, plus the economic effects, feasibility and conflicting interests of, for example, agriculturalists and conservationists, can we suggest methods of increasing production from grasslands both on an extensive and intensive scale?

Intensive grazing management. We have demonstrated selectivity in grazing of different floral species, growth stage of the plant or part of the plant. Different faunal species and different age groups of animals have been shown to select different floral species and even different sexes and individuals may show varying preferences. Thus it would seem logical to plant a ley relatively rich in floral species, of relevant types for the stock that will graze it, and then to graze such a ley by a fair variety of different species, age groups and sexes of animals.

Let us take a test area in South-East England to see how an approach of this type could be integrated with increasing production. South-East England is an area of very high amenity value; every inch of land is tended and watched with care by various sections of the community with varying interests. Agricultural interest at present is orientated towards digging up the hedges and making vast fields out of the patchwork countryside. This, needless to say, destroys the traditional pattern of the countryside and reduces its amenity value. Can we suggest a management policy that will increase production on such areas, while keeping the wealth of amenity value?

Proctor et al. (1950) have suggested that maximum production can be gained off a ley or permanent pasture by paddock grazing in rotation. At the end of the grazing the ley must be close grazed, as this encourages growth. The grass should be topped by a mower just before the end of each grazing season so that the defaecation tufts of grass are removed from the presence of the faeces and the animals will consume them. Thereafter, in order to prevent reduction in utilisation due to faecal contamination, the ley should be grown for hay or silage before being grazed again.

An appropriate rich ley should be planted and, in addition, different species grazed. There is some evidence that mixed grazing (in agricultural parlance this indicates cattle and sheep) increases production of a ley. Hamilton & Bath (1970) showed that production from a grass ley increased in 3 out of 4 years with mixed grazing at 3 different stocking rates, compared with sheep only or cattle only. Culpin et al. (1964) also found that production was increased when cattle and sheep were grazed together on ley pastures. Leaver (1970) found that production was increased when heifers and calves were grazed as compared with either one or the other. In all these experiment higher stocking rates increased production, as the two species, age groups or sexes grazing allowed for differences in food selection, thereby reducing the amount of wastage.

In addition, a mixed grazing scheme would allow for the inclusion of amenity animals (such as donkeys and horses) in a useful way. In South-East England these are numerous and the demand for them is increasing.

Therefore, the following policy is suggested: to start with, because the tops of grass contain the highest percentage of digestible matter and protein, animals which are required to grow quickly, are lactating or are in the last stages of pregnancy should be rotated first on the ley (e.g. dairy cattle, quick-growing young beef animals, mares and foals, ewes in late pregnancy and early after parturition). Next, animals requiring maintenance only and/or able to close graze should be placed on the ley. These would be breeding males, riding horses and ponies, dairy herd replacements and dry cows and, perhaps, geese.

The problem is to work out equivalences to ensure the correct
number of each species. Although there are detailed figures on intake in
a grazing situation in cattle and sheep, there are none for the less
conventional species such as horses and geese. However, with some
thought these can be calculated as a guide line (a calculation along
these lines done for me by Mr Bob Large of the Grassland Research
Institute, Hurley, indicated that a similar area could support one cow or
185 geese!)[2]

Such a scheme would not be suitable for all areas and the species
managed would, of course, depend largely on the area. In the South-East
of England, an area of high amenity use and where production must
also be high, perhaps such an approach would help to reduce the
conflict between agriculturalists and conservationists. Small fields and a
varied countryside would be necessary to such management and
production could be very high if correctly managed.

Extensive grazing systems. In this context we will take the upland
moorland areas of Scotland which present difficulties for farming and
conservation alike. This marginal land has traditionally been grazed by
sheep and as a result has suffered considerable deterioration in the last
100 years or so (Hunter 1960; Ratcliffe 1965).

The germinating woodland and *Erica* spp. are systematically destroyed
by the sheep and a system of management involving too frequent
burning. Thus there is a loss of primary production, reduction in soil
fertility and the formation of screes (Ratcliffe 1965). These
developments are clearly detrimental to wildlife (including the grouse
and red deer which account for a substantial amount of the income
from such areas) and result in further loss of production in sheep. As a
result sheep farming in the hills has today become generally unprofitable
and is maintained largely by government subsidy.

Large scale improvement of the hills by ploughing and reseeding is in
general neither successful nor economically viable except on easily
accessible lower slopes. In addition, as Tribe (1950b) points out, a
mixed flora must be maintained if year round subsistence is required.
Similarly, Dale (1965) in Snowdonia showed that a varied flora allows
for better production in bad years as well as good.

Mixed species grazing of agricultural animals can increase production
on an extensive scale as well. This has been shown by Peart (1962) who
found that the addition of cattle to traditional sheep hill pastures
increased production from 26lb/acre to 54lb/acre! Clarke (1963) found
that not only was total production increased when cattle were

introduced to traditional sheep grazing, but that the sheep also benefitted. It therefore seems logical to extend such thinking to include, for example, red deer, goats and ponies, as well as sheep and cattle. The hardy highland ponies of various breeds in Britain have hardly been considered in this way although, irrespective of their possible use for meat, there is a growing market in America and the Continent for such animals for amenity use.

Thus a balanced number of cattle, sheep, ponies and (possibly in some areas) goats could be grazed throughout the year, but with the stocking rate increased substantially in the summer for sufficient grazing pressure to encourage the more palatable species.

Very steep slopes and other areas difficult to graze should be fenced off to allow the regeneration of natural scrub and woodland. This could include the planting of some hardwoods and climax oaks, as well as the planting of some conifers as a nurse crop and to give a shorter term return. The wholesale planting of conifers would not be a recommended policy, however, since the central idea of such a programme would be the rebuilding of soil fertility, which most conifers do not enhance. Once these areas of scrub and woodland are well established, grazing could be allowed within them, and thereafter other areas fenced and allowed to regenerate. In this way it is suggested that it would be possible to gradually increase soil fertility again while maintaining production, and without large capital investment in draining, liming and so on.

Such grazing schemes are of course more complex and require more skill in their management than the more conventional management schemes but it may be that we should begin to develop these skills so that our grandchildren will inherit a more productive and less polluted world involving the sensible use of renewable resources.

[1]First published in The Management of Ungulates and its Relevance to Management. IUCN 1974. Reproduced here with permission of IUCN.

[2]Over the last four years I have been trying this type of management on a small farm in Sussex of 30 acres. Our carrying capacity is still rising and now stands at 1 bovine unit per acre, but without the use of any artificial fertilizers, pesticides or herbicides. This is slightly better or at least equivalent to the various neighbours carrying capacity on their rye grass leys carrying dairy cows which are fertilized at a rate of up to 4 cwt per acre per year.

CHAPTER 15
STOCKMANSHIP

Good stockmanship is more than just good management of stock. It entails the ability of the man or woman looking after and handling the stock to detect any change in behaviour or health. This will enable the animals to be more productive or aid their welfare. Early diagnosis of disease, oestrus or other factors, which are essential to good management, are normally behavioural. Thus the good stockman must have a substantial knowledge of his stock (and if possible the individuals within his care) and their behaviour. He must also be prepared to act as a result of such knowledge even if this is to the detriment of his own personal life (for example, being able to detect imminent parturition—and remain with the animal during the night if necessary).

Stockmanship frequently controls whether an enterprise will be a success or failure. Most farmers emphasise the importance of stockmanship.

Some examples of the importance of the stockman in agricultural enterprises have been quoted to me. Thus Hayward Bros indicate that about half the success of their pig enterprise (which is considerable) is due to the stockman. Paton (veal producer, Spaxton, Somerset) says that not only can he tell what stockman has looked after the veal calf when it is alive from its performance but also from the carcass quality when dead. Soutar (Farm Buildings Investigation Unit, Aberdeen) says that it is possible to tell the character of the stockman from the behaviour of his pigs! In dairy herds a bad stockman who hurries the dairy cattle will cause problems with bullying at gates and poaching of the ground, and nervous kicking in the milking parlour (Hooklands Estates). Stockmanship affects production directly in some cases; for example, Baryshnikov & Kokorina (1959) found that milk was not let down to a strange stockman.

Stockmanship clearly involves the relationship between the husbandry man and his animals. The type of this relationship indicates whether the man is a good or a bad stockman. In the case of a good stockman the animals do well and production is high, whereas a bad stockman can reduce profits to nil, although he apparently does all the jobs that are expected of him. The classic example of this effect is illustrated by the experience of a Hertfordshire farmer, Mr Patterson, who has classified his stockmen in terms of the yield they obtain from the dairy herds he owns. As a result he can move the stockmen around from farm to farm, on each one of which he has a similar routine and dairy herd, and finds that a good stockman will obtain up to a 20% increase in milk yield over a bad one on the same farm. The factors that contribute to this difference are obscure, but they are apparently not related to routine and time spent doing the various operations during the milking. Thus the effect appears not to be related to the mechanical efficiency of the man doing the job but rather in other, to date unmeasurable, variables, which could be summarised perhaps as the attitude of the man towards the stock (this may be reflected in how he handles them, whether he talks to them and so on, and his ability to observe them) and their response to him (personal comment Dr Lloyd, University of Reading). Similarly, the Milk Marketing Board find that AI inseminators tend to have individual characteristic percentages of success.

Nowhere are the effects of stockmanship more noticeable than in the stables. Here, incorrect handling by inexperienced or ignorant people can result in horses becoming virtually unusable through the development of attacks made on humans, nervousness or extreme reactivity to environmental change. The same applies in animal training where it is the personality of the trainer and his ability to predict what the animal will do next and to forestall it, or modify its response, that is important. In many ways, this rests on the ability to communicate with the animal. Similarly, successful circus trainers have two particular principles:
(i) to use positive reinforcement rather than negative wherever possible and
(ii) to teach the animal to do something it would naturally do, perhaps to an exaggerated extent. This is easier than to teach the animal to do something alien, and implies that one must have a knowledge of what the animal would naturally do, as well as its likely response to different situations.

As units become larger and larger it is clear that it becomes more difficult to know animals individually, and less easy to regard the animals as individual living creatures. As a result, the ability to foresee their needs declines, whilst the motivation to recognise and attend to symptoms of environmental difficulties is reduced if the man is in no way emotionally involved.

If men and women do find it more difficult to become good stockmen in very large intensive units, one answer would be to split up the units and make individuals responsible for all the requirements of a

specific number of animals. Most farmers will agree that this is the
most profitable way of operating dairy herds. Alternatively, one might
try to reduce further the importance of stockmen. This is likely to be
difficult even with totally automated feeding. Animals will still have to
be checked over for injuries and symptoms of disease periodically, and
these are more difficult for strangers to see than a man who knows the
normal behaviour and performance of his animals. One of the major
reasons given for not installing fully automatic feeding systems in some
houses is that the stockman must visit each pen at least every day if
the animals are manually fed. As a result each animal is checked daily,
however cursorily.

To illustrate the fact that the method of handling has considerable
effects on the responses of the animal, let me quote Rosen (1958) and
Bernstein (1957), who showed that 'gentling' (gentle stroking of the
animal) before weaning in rats and mice reduced emotional responses
and increased learning ability. Weininger (1956) found that gentled rats
which were bandaged in gauze after weaning produced fewer
physiological effects of stress than the ungentled ones. Denenberg (1964)
in his review emphasises the difference in response as a result of
'gentling' the rats and mice and 'handling', which appears to be a
rougher, more traumatic treatment. Gentling appears to decrease
physiological stress responses, whereas handling increases it. There is
anecdotal evidence to support this claim in domestic animals such as
horses and dogs, where early 'gentling' increases the ease of training
whereas rough treatment by humans results sometimes in the
development of behavioural pathologies and difficulties with training.

It is not suggested here that all animals in the farm situation must be
stroked for several minutes a day. However, there is evidence that the
method of handling can affect physiological and behavioural responses
of the animal, some of which may well interfere with production.

Further, it is possible that the cowman who has attended a cow in
labour and throughout life may acquire a special relationship to the
cow and can be regarded by it as, in some sense, its calf. Thus the
conditioning stimuli for the let down of milk may involve, for example,
the movement of buckets, the total environment of the milking parlour
and, in addition, the presence of the familiar man or woman who will
actually place the machine on her teats.

At all events, change in the milking staff almost always produces a
drop in milk yield. Clearly the milk yield can be re-established after the
new man has been present for some time, but the best responses from

the cows are obtained from herds where the same man has been present for long periods of time.

Some people maintain that women are best for certain jobs, such as raising calves. The raising of calves in large numbers is a particularly skilled process involving very rapid recognition of any symptom of disease or unrest in the calves and considerable patience. Stockmanship presents problems which will be of particular importance as unit size and intensity increases. Interdisciplinary research involving sociologists, ethologists and agriculturalists is urgently needed here.

CHAPTER 16
HOUSING

A full discussion of housing and its relationship to behaviour could be the subject of a complete report on its own. However, there are several aspects of the problem which can fruitfully be included here.

It is often the case that, as Ross (1960) states when considering swine housing—"too often the buildings are designed for the comfort and well-being of the caretaker rather than the pigs". If building design is to take into account the comfort of the animals, then clearly one must have some measure of this. An effort to measure 'comfort' in terms of time spent lying down has been made (Ministry of Agriculture, Fisheries and Food 1965). This appears to depend upon too many variables, and other methods are needed. Unless we have detailed knowledge of the normal behaviour of the animal we cannot assess this at any level. Even then measurements of one or two behaviours, such as lying and their divergence from the normal may further confuse the assessment. For example, there is a suggestion that animals will lie and sleep more, not only when they are 'comfortable', but in order to switch attention or switch off from an unacceptable environment—lying here would represent a 'cut off' behaviour.

We will therefore confine the discussion in this chapter to aspects of housing which have been considered particularly important in causing or helping to cause the behavioural problems we have discussed in the previous chapters.

Behaviour at high temperatures. Most animal housing in Britain is designed chiefly to increase the winter temperature. As a result temperatures may become excessive in summer. Under these conditions animals may develop abnormal behaviour. For example, one of the most common abnormal behaviours in pigs in the summer in Britain is dung wallowing. The animals will dung all over the pen, then wallow in it, presumably cooling themselves in this way. Alternatively, they may play with their waterers and indeed even fall asleep with their snouts pressed on the release mechanism, thus creating a river to lie in.

Pigs will disperse themselves, if given the opportunity, in a hot environment (Fraser 1968b). Nevertheless, they are stocked at the same densities in their pens in the summer as in the winter. It is possible that self-operated fans or water spouts could be provided cheaply in pig houses which would take the place of full air conditioning. This might well reduce aggression also by being occupationally therapeutic. It has been shown experimentally that pigs quickly learn to control their own environmental temperature (Baldwin & Ingram 1967) using operant conditioning.

Poultry develop feather pecking and cannibalistic habits in high temperatures and crowded conditions (Ferguson 1968). Proper ventilation, or allowing access to the exterior during the hotter months, helps to reduce this problem (Mayhew Chickens Ltd, Uckfield).

Sheep and cattle are usually out of doors in the hotter months of the year and so able to make behavioural adaptations to increased temperature (e.g. shade seeking). Store cattle houses today are, in any case, not usually totally enclosed so that ventilation is good and temperatures rarely climb too high. However, in the future it can be foreseen that heat stress in an enclosed building may become a problem in beef rearing units in particular. 'Controlled environment' calf rearing houses are popular, but Mitchell & Broadbent (1973) found that calf health and performance was not substantially improved in them when compared with an unheated 'climatic house' even in the North-East of Scotland.

Behaviour at low temperatures. It is usually assumed that low temperatures reduce productivity and modern agricultural housing is carefully designed to cut down heat loss. In general, however, adult domestic animals are able to adapt well to quite severe winter temperatures. Indeed, at the University of Guelph farm where the temperature may drop to -40°C in the winter, sows are kept outside with only a shelter. The main problem encountered under such a management system is to stop the water freezing in the drinkers, not the survival of the sows!

Davis et al. (1967) found that calves were healthier and heavier at 120 days old when raised in outside pens than their controls in conventional indoor housing. This test continued for 2 years and the outdoor calves were found to have less coccidiosis, greater weight gain and less diarrhoea than the indoor calves.

Adaptation to low temperatures in confined environments is probably less difficult than adaptation to high temperatures. Usually animals crowd together and rapidly raise the temperature of their immediate surroundings (particularly pigs). The addition of bedding obviously helps the animals to keep warm. The modern tendency, however, is to omit bedding and provide slats over slurry tanks. These are normally wet and have a damp cold draught from below.

Sheep are able to withstand inclement weather in the hills involving not only low temperatures but also wind and rain, by finding the most sheltered areas. In winter shelter is of so great an importance to hill sheep that intake may well be reduced, since the animals will not leave their place of refuge in order to feed (Hill Farming Research Organization).

Ventilation. There has been a considerable body of work on the ventilation requirements of housed animals, and it has been shown that inadequate ventilation will reduce weight gain in beef (Hanson & Mangold 1960) and other animals. Deaths have been reported in pigs from gases when underfloor sewage tanks were being pumped out. Bad ventilation is one of the factors that affect tail biting and other abnormal behaviours (page 9).

Light. The induction of oestrus in ewes by changing the day length is a common practice today. However, the percentage of conceptions that result from such practices is low (R. Wilson, Hurley Grassland Research Institute). It remains to be seen whether this is related to infertility in oestrus on the part of the female or to seasonal changes in male fertility.

Fertility was also shown to be reduced in a herd of dairy cattle kept in reduced light intensity through the winter by Deas (1970). Reduced light intensity is widely used in pig and poultry units to reduce tail biting and feather pecking respectively. Shultze et al. (1960), however, found that this practice reduced growth rate in chickens. They suggest that this is because intake is reduced. It may have a similar effect on pigs, although Hanson & Mangold (1960) found little influence of reduced lighting on growth rate in porkers. It seems likely that a reduced lighting regime for sows might well affect the onset of oestrus.

Reduced lighting in pig, poultry and calf units is considered to reduce activity and thereby increase growth. No controlled measures have been made here, however.

On the other hand, Shultze et al. (1960) showed that keeping chickens on a 24 hour lighting schedule increased growth rate. A number of birds, however, had enlarged and flattened corneas and, between the 4th and 5th weeks, they were particularly prone to enteritis and encephalomalacia, indicating possible physiological stress (see page 4 et seq.).

Flooring. The type of flooring the animal is on has a considerable effect on behaviour. If slats are wide and the gaps between them small, they appear to be quite successful for cattle and pigs, although concrete ones are said to be better than wood (Soutar, Farm Buildings Investigation Unit, Aberdeen). Slippery slats result in damaged muscles and joints, and their slipperiness may discourage the animals from lying down. The absence of bedding is one of the contributing factors to the development of tail biting in the pig (Gadd 1967 and personal comments from various farmers). The type of flooring can have direct effects on production. Verstegen & van der Hel (1974), for example, found that weight gains in growing pigs on asphalt and straw bedded floors were significantly higher than on concrete slats.

Bedding. Against the economic reasons for avoiding straw bedding one should offset the apparently therapeutic effects of straw in pig units and cattle housing. For example, it has been suggested by various farmers, and the author, that straw can act as a 'toy' for occupational therapy. One of the objections to straw is that it clogs drains. However, in pig and cattle units it is not difficult to restrict bedding to a specific area by using container bars, thus avoiding such clogging. The behavioural effects of the presence or absence of bedding deserve further investigation. Its potential importance is suggested by the fact that animals may make attempts to manufacture their own bedding; thus, in Guelph University Farm pig unit, the weaners were lifting meal out of the hoppers in order to use it as bedding. Swedish work (Ekesbo 1974) indicates that health is affected by lack of bedding. Here mastitis was reduced in dairy herds bedded on straw as compared with bare floor.

Sow stalls and cubicles for cattle. If animals are placed in cubicles when young, they quickly become accustomed to them (Leaver & Yarrow 1970). However, unaccustomed cattle may well lie in the passageways and dunging gutters until they get used to the cubicles. Whether cubicles are indeed more economic in that they can cut down the total amount of space allowance for each animal remains a matter of dispute. Stalls and cubicles are often made very short in order to ensure that the animal occupying it defaecates in the drainage channel. . An alternative approach is to provide larger, more comfortable stalls but with an electrified wire above the back of the stall which shocks the cow as it arches its back prior to defaecation, thus ensuring that she moves back. These are in use in several dairy units where the animals are stalled with apparent success. They could equally well be used in sow stalls, thereby allowing the animals a larger stall for more comfortable lying. Such techniques could make the provision of bedding within the stall or cubicle more economic since it would not become soiled.

The long term effects of sow stalls have yet to be determined. One major problem at the moment appears to be that they make the recognition of oestrus by the stockman difficult.

Windows. The importance of environmental stimuli has been emphasised throughout this report. One way of providing varying visual stimuli is the use of low level windows in pig houses. Arguments against this idea have been posed in terms of increasing the cost of building and heat loss. However, the added gain in behavioural adaptation might offset such expense. One of the best and most successful pig houses I saw was an open ended 'Suffolk' type pig house, although here there must be much greater heat loss than through a series of double glazed windows.

Self-feeding and zero grazing. Self-feeding of silage is a useful way to cut down labour costs. However, in crowded houses adequate silage face must be provided: thus Mr Delaney (Hookland Estates) found that he had to increase the amount of feeding face from 6 inches to 9 inches/cow, since some animals were pushing others off and as a result intake was reduced in some animals. This is likely to be a limiting factor on the degree of crowding which is tolerable. In zero grazing systems it has been found that 3 out of 70 cows did not obtain any grass from the trough as they were too 'timid' to try. The answer to these problems may be conditioning the animals to enter individual stalls to obtain their own personal rations. Gadgets involving the wearing of a small transducer around the cow's neck which unlocks the appropriate stall are now being tested at the National Institute for Research in Dairying (England).

Very many of the behavioural problems outlined in this monograph are the result of the housing having been designed for the stockman. In so far as the animals have been considered, judgements on their requirements (other than food) have been anthropocentric. We are now accumulating knowledge slowly on the effects of different housing designs on behaviour and production of the different species, and it is possible that in the future building design for stock will be more successful.

There are other neglected factors also in farm building design which it is becoming imperative to consider. One of these is that farm buildings, including those for stock, are the Cinderella of buildings, as it were, and are usually built for the least possible cost our of the cheapest materials. The effects of this policy is firstly a reluctance to consider the animals' psychological and physical requirements since this usually cost more money (for example, the fitting of windows into stock buildings). The other important effect of this 'scrimp and save' attitude is a lack of consideration of aesthetics and the effect on the countryside. To date the planning on farm building design has been very lax or totally lacking. It is difficult to see why this should be the case but public opinion is awakening slowly to the growing factory effect of modern farm buildings in the rural landscape.

One solution to these problems is to reduce the amount of meat the human population eats. Should we have cheap meat at the expense of a) the animals so raised and b) the industrialisation of the countryside? Reducing the meat eaten and therefore the further spread of intensive livestock husbandry is easily done by raising the price. The added return to the farmer could then be controlled in such a way that we could have better farm building design for the stock, the countryside and the farmer and stockman. The added advantage here would be that less livestock would require less food, and the land so liberated from growing stock food could be used for growing food for human consumption direct, thus making the nation more self-sufficient in food stuffs.

106

CHAPTER 17
DISCUSSION AND CONCLUSIONS

The main behavioural problems as the result of modern husbandry practices have been identified in this monograph. Pertinent information from work with other species and from other disciplines has been reviewed to try to clarify the causes of such behavioural problems in farm livestock. As a result, it is quite clear that in many respects we know far less behaviourally about these species than about animals such as the stickleback, rat and gull despite the fact that humans have lived with these domestic animals for around 4,000 years! Much of the information that is available on the behaviour of farm livestock is still anecdotal and relies heavily on the experiénce of those tending the stock.

There is sufficient information, however, to stress that most of the behavioural problems of agricultural animals are not dependent on one particular environmental factor (such as, for example, dietary deficiency or group size) since they may occur under many different environmental conditions. Until recently the cause and therefore the treatment of behavioural disorders (such as with medical disorders) in man and animals has tended to be related to particular pathologies or to one environmental factor. Thus cannibalism in chickens has been related to dietary deficiency (Siren 1963b), and human schizophrenia to brain malfunction or lesion (e.g. Fish 1962). A more helpful approach, particularly with preventative treatment, is to consider the total environment of the patient (e.g. Laing & Esterson 1964 in human schizophrenia). In many cases it is likely to be a summation of a number of different causal factors, in both the external and internal environment, that gives rise to the problem. This is also concluded by Hughes & Duncan (1972) in their review of cannibalism in fowls.

To illustrate this point further, the accompanying table is given which summarises some of the information in this monograph. For the sake of clarity, references are not given here but will be found in the text. It links the main behavioural problems with the environmental factors reported to affect it. The idea of summative causes makes both the prevention and treatment of such problems much more complex, and has proved difficult to accept. For example, Gadd (1967), although he admitted that type of food, ventilation, crowding and confinement can effect tail biting in pigs, then concludes that it can be cured by liquid feeding!

Tracing the cause of various behavioural abnormalities on the farm has proved extremely difficult for another reason. This is that the behavioural work that has been done on farm livestock rarely relates changes in behaviour to a single variable, but usually confounds many. This is largely because workers have been primarily interested in production indices rather than behavioural.

This report raises more questions than it answers. However, it does point out the areas requiring further research. Those that are of particular importance are:

(1) Changes in social organization as a result of intensification, and an increase in aggression.

(2) Problems of reproductive behaviour, including such things as (a) infertility, (b) oestral recognition, (c) social effects on ovulation and conception, (d) deficient maternal behaviour and (e) decrease in male libido.

(3) The causes and ontogeny of stereotypies such as inter-sucking and crib biting.

(4) The effect of various environmental factors on food intake.

(5) The grazing behaviour and how environmental factors affect this.

(6) The effect of changes in lighting and temperature on behaviour.

(7) The effect of environmental changes and complex versus 'dull' environments on behaviour.

(8) The effect of isolation and social facilitation on behaviour and performance.

There are in addition two particularly fundamental areas where the results of proper behavioural research could help towards an understanding of several of the problems above. These are:

A. There is a need for proper quantitative studies on the social organization and behaviour of farm animals under free-range conditions. This would provide a norm from which degrees of deviation could be measured and assessed. Without such a baseline, abnormal behaviour cannot be properly identified.

B. Behavioural measures of stress are now badly needed. Veterinary surgeons are being asked to implement codes of practice without any agreed measures of this type. In addition, situations and environments which give rise to physiological responses of stress will certainly interfere with production as well as welfare. It is essential, therefore, (since it is impractical to suggest that physiological measures of stress must be frequently assessed in the farm situation) that some reliable measures or stress are found. One first step would be to correlate measures such as posture, responsiveness and vocalization with agreed physiological indices such as corticosteroid levels. Such work should be carried out on all species of farm animals which are now being placed in intensive units.

It is the writer's belief that if the behaviour of the animals is clearly understood and catered for, then behavioural therapies, such as the use of occupational therapy and conditioning hand in hand with suitable environmental design, will be sufficient to eliminate the majority of these problems or at least reduce them to negligible levels. However, there are other curative or preventative measures currently in use. The most common of these is the use of drugs and surgery. Again, these have been discussed in detail in the appropriate chapters, but a brief summary is given here.

The use of drugs and hormones. Drugs such as tranquillizers can be very useful in some instances, and for particular manipulations (e.g. when transporting, marketing and manipulating large animals). Nevertheless, the large scale and constant use of drugs to combat these problems is to be avoided as much as possible. If the behaviour of the animal is understood and taken into account, the sustained use of tranquillizers will never be necessary.

The use of hormones to synchronize oestrus (in sheep and cattle, for example, or to induce ovulation in horses) is already widespread, although the number of conceptions from induced oestrus is low. Behavioural solutions may prove to be more long lasting, more efficient and less expensive. Thus, for example, the presence of a male or of olfactory stimuli from the male might well render the use of gonadotrophins unnecessary in the horse. Various animals, particularly beef animals, are treated with steroids which are reputed to have anabolic effects. A recent investigation of the anabolic effects of androstendione in young ruminants indicates that it has little application in meat production (Johnson & Wilkinson 1974). Should further work indicate that treatment with hormones or other drugs does have anabolic effects, it is essential that the behavioural effects of such treatment be investigated.

The widescale use of antibiotics in agriculture has recently given rise to adverse public opinion and questioning and has been widely condemned (e.g. Swann Report 1969). The fact that such measures have to be taken to contain disease in intensive units indicates that such environments are inadequate from the animals' point of view. It is reasonable to suggest that common outbreaks of disease may be attributed to prolonged physiological responses to stress. Again, proper understanding of behaviour and environmental demands of the animals may well render such preventative treatment unnecessary. Similarly, the outbreak of other herd diseases may be controlled through similar pathways. Thus the study of the occurrence of herd diseases is likely to prove a particularly profitable approach. The first attempts at looking at this are being made in Sweden (Ekesbo 1974).

The use of Surgery. The use of surgery has grown over the last 50 years and is so accepted by farmers that many of them believe, for example, that piglets cannot survive without having their canine teeth cut or their tails docked. There are clearly some surgical techniques that can be of very great use in an intensive system. An example of this is dehorning of yarded cattle. Provided this is done properly and in the young animal, there can be no objection. Dehorning reduces harmful aggressive exchanges and makes the animals easier to handle and therefore they can be housed more densely. The docking of the tail in sheep is another operation which is necessary to prevent infection from accumulated dried faeces. Docking the tail in pigs, however, appears occasionally to help prevent tail biting but just as often results in ear biting instead. Does one then cut off the ears? A more sensible approach is to regard tail biting as a symptom of bad environmental design and change the latter.

The uses and abuses of castration have been pointed out (page 60). The original reason for castration in the majority of cases was to control the number of breeding males, obviously of great importance in free-ranging animals. However, this wholesale castration needs reassessing where there are modern purpose-built intensive units. It is possible that the extirpation of various glands might reduce, for example, intra-group fighting or boar taint to meat while retaining the anabolic effects of testosterone but, again, a proper understanding of the role of scent signals might lead to simpler and cheaper behavioural solutions.

Debeaking of chickens and turkeys is a subject that has received some attention recently. Should it be essential to debeak to reduce cannibalism, then a thorough look at the whole environment should be taken to assess the possible causes. The loss through cannibalism may be considerable, but the loss in production due to sub-optimal conditions that necessitate the debeaking is likely to be greater.

Animals respond well to routine, and profound environmental change is traumatic (page 28 et seq.). Thus animals that are accustomed to, for example, controlled environmental temperatures and restricted environments when released to pasture would find this very traumatic and, it could be argued, this is then 'cruel'. An example of the effects of this is found in battery-reared hens which are released onto free range. The environment, as well as being totally alien to them, is fraught with problems (such as having to search for food, compete with other hens and so on). Such problems they have never had to face before. This experience can well result in death from no clear cause. A similar response is shown in the capture of wild animals. This has been related to 'overstraining muscle disease' which appears to be a response to extreme stress.

There are limits to how restricted and confined the animals can be successfully reared and kept, and there is little need to overstep this mark. Thus all animals should have sufficient space to be able to groom themselves, lie and stand at will and carry out as many normal activities as practical (Codes of Practice 1971 and Brambell 1965). It would be foolish not to allow such behaviour since its restriction results in crib biting and skin diseases. In the young, inadequate provision for suckling results in inter-suckling which can continue into adulthood.

This monograph has discussed the various behavioural problems encountered in modern husbandry, and has pointed out the main areas requiring research. Within a practical context such an academic approach is unsatisfactory. As an applied ethologist one must have more pragmatic solutions wherever possible, pending confirmation or dismissal from proper behavioural studies. The biological approach to this is somewhat different from that of the agriculturalist or veterinarian. It is to consider that the animals have evolved to cope with their own environment, forest or plain, social or solitary and, therefore, will be most successful (biologically speaking) within this type of environment. Thus the nearer to this that can be approached in the husbandry system (whether this is by substitution or not) the more successful the animal is likely to be in terms of both health and production. It is argued, of course, that it is unlikely that basic behavioural modalities such as social organization and habitat choice will have changed greatly during the course of domestication since neuro-anatomy and physiology have not.

We cannot end without some mention of the welfare and ethical aspects implicit in keeping animals in modern intensive systems. In favour of intensive husbandry it has been argued that animals so housed are away from the vagaries of the weather and readily supplied with food, and therefore suffer less than their cousins under more natural conditions. This is illustrated, the exponents say, by the continued productivity and increasing growth rate of animals in intensive systems. Ewbank (1969b) discusses these assumptions and concludes that such arguments are usually spurious. This monograph has also contributed to this debate by pointing out where we have evidence of behavioural problems, which might be taken to be one index of some degree of suffering, and the causal factors that give rise to them. More recently, a philosopher, Peter Singer (1976), has discussed such arguments and the ethics of using animals entirely for the benefit of man with logical and philosophical erudition. This is uncharacteristic of such emotionally charged issues as 'factory farming' and 'vivisection' and is therefore most welcome and enlightening. The crux of the whole argument lies in whether or not one believes that there is a fundamental difference in the amount and type of suffering that can be undergone by *homo sapiens* and other mammals. Before we can answer this question with certainty in all cases we need more knowledge, particularly behavioural knowledge, of both the other mammals and man. However, as we have seen, an examination of the behavioural problems of some mammals indicates that under certain conditions these animals show behavioural problems which are in many ways similar to those suffered by man.

This is no place to go into the ethics of animal husbandry in detail, but I do believe that it is the responsibility of everyone in any way connected with the use of animals to examine carefully their own ethical position on such issues rather than close their eyes to it. This is, of course, particularly the case for veterinarians and others pledged to work to reduce animal suffering.

Table 1   A comparison of various behaviours in wild and domestic species which are now used in agriculture

SPECIES	HABITAT	GROUP STRUCTURE				TERRITORIALITY		GROUP MOVEMENTS		DOMINANCE		
		Pair living	Family groups	Several family groups	Large herds	Defence of specific area	Defence of proximity to individual	Home range	Migratory	Rigid hierarchy	Less rigid	Not important
Bos												
Auroch	Edge of forest & plain		★	★		?	?	?		?	?	?
Syncerus caffer	Savannah		★	★			★	★	★	★?	★?	?
Bison	Prairie & forest		★	★				★	★	★		?
Feral Cattle	Edge of forest & plain		★	★		★ (older males)		★		?	★?	★
Ovis												
Mouflon	Mountains & scrub	?	?	?			★?	★			?	
Mountain sheep	Mountains & scrub	★	★	★			★	★	★	★?	★?	
Soay sheep	Hills/moorland	★	★	★		?	★	★		★	★	
Hill and domestic sheep	Hills/moorland	★	★	★		?	★	★		?	★?	★?
Capra												
Wild (feral) goats	Desert to forests	★	★				★	★			★?	
Domestic goats	Desert to forests	★	★				★	★			★	
Equus												
E. zebra (spp)	Plains	★	★	★			★	★	★		★?	★
E. caballus (feral)	Plains & scrub	★	★	?			★	★	? ★	★	★	★?
Domestic		★					★	★		★	★	
Sus scofa												
Wild boar	Forests	★	★	★		? ★	★				★?	
Feral pig	Forests	★	★	★		? ★	★				★	
Domestic pig	Forests									★	★	
Gallus (Fowl)												
Gallus gallus (red jungle fowl)	Forests & vleis	★	★				★	★		?	★	
Gallus domesticus (domestic hen)		★	★				★	★		★	★	
Turkeys												
Meleagris gallopovo (wild)	Mountains/forests	?	?			?	?	?		?		
Domestic		★	★					★				
Anatidae												
Anatinae wild ducks	Marshes & near water	★	★	★	★			★	★	★			★
Domestic ducks		★	★	★	★			★	★	★			★
Anserinae wild geese	Marshes & grazing land	★	★	★	★			★	★	★			★
Domestic geese		★	★	★	★			★	★	★			★

Tendency to follow	Tendency to group	Contact seeking				FOOD PREFERENCES			SOURCES
		Tactile	Visual	Olfactory	Auditory	Grazers	Browsers	Scavengers & others	
?	?	?							Cole 1961
★★	★★	★★	★★	★★	★★	★★	★		Sinclair 1974 (wild) pers. obs.
★★	★★		★★	★★	★★	★★	★★		McHugh 1958 pers. obs.
★★	★★	★★	★★	★★	★★	★★	★		Kilgour 19 (wild)
									Schloeth 1961 pers. obs.
★★★?		?				★★ ?	★★ ?		Kiley pers. obs. (zoos)
★★★	★★★?	★	★★★	★★	★	★★	★		Geist 1971
★★★	★★★	★	★★★	★★	★	★★	★		Grubb & Jewell 1974
★★★		★	★★★	★★	★	★★	★		Hunter 1962
★	★	★	★★	★★	★	★	★★		Geist 1965 (wild)
★	★	★	★★	★★	★	★	★★		Pers. obs.
★★	★	★	★★★	★★	★	★★	★		Klingel 1974 pers. obs. (wild)
★★	★	★	★★★	★★	★	★★	★★		Tyler 1972 pers. obs.
★★	★	★	★★★	★★	★	★★	★★		Pers. obs.
★	★	★★★	★	★★★	★★★	★	★	★★	Gundlack 1958 (wild) pers. obs.
★	★	★★★	★	★★★	★★★	★	★	★★	Fradrich 1974
		★★★	★	★★★	★★★	★	★	★★	Fradrich 1974
									Ewbank 1961 pers. obs.
★★	★★	★	★★★	★ ?	★★	★	★	★★	Krujit 1964 (feral/caged)
★★	★★	★	★★★	★	★★	★	★	★★	Guhl 1953
									Wood-Gush 1971
?	?	★	★★★	★ ?	★★	★ ?	★ ?	★★ ?	Anecdotal
★★	★★	★	★★★	★	★★	★	★	★★	Hale & Schein 1962
★★★	★★★	★★ ?	★★ ?	?	★★?	★	★	★★	Weismann 196 (wild)
★★★	★★★	★★	★★	?	★★	★	★	★★	Collias 1962 pers. obs
★★★	★★★	★★	★★	?	★★★	★★	★	★	Kear 1966 (wild)
★★★	★★★	★★	★★		★★★	★★	★	★	Pers. obs.

of the physical environment

Table II ENVIRONMENTAL CONDITIONS

	Behavioural changes	Sudden environmental change	Restriction (tied)	Insufficient space/individual	Restricted rumination	Absence of key stimuli	Absence of bedding	High temperature	Low temperature
	Increase in aggression	Pigs, Cattle, Sheep, Horses, Chickens & Turkeys		Pigs, Cattle, Sheep, Horses, Chickens & Turkeys	Cattle		Pigs, Cattle	Pigs, Cattle, Sheep, Horses, Chickens & Turkeys	
	Increase in activity	Pigs, Cattle, Sheep, Horses, Chickens & Turkeys		Pigs, Cattle, Horses, Chickens & Turkeys	Cattle		Pigs, Cattle		Pigs, Cattle, Sheep, Horses
	Smothering	Chickens & Turkeys		Chickens & Turkeys				Chickens & Turkeys	
Problems of reproduction	Reduction of fertility			Cattle, Sheep		Pigs, Cattle, Horses		Pigs, Cattle	Pigs, Cattle
Problems of reproduction	Non-recognition of oestrus			Pigs, Sheep					
Problems of reproduction	Oestral behaviour causing disturbance	Pigs, Cattle	Cattle	Pigs, Cattle, Sheep					
Problems of reproduction	Insufficient libido		Pigs, Cattle	Pigs, Cattle		Pigs, Cattle, Sheep			
Problems of reproduction	Abnormal egg laying	Chickens & Turkeys		Chickens & Turkeys			Chickens & Turkeys	Chickens & Turkeys	
Problems of maternal behaviour & young	Mother ignoring young	Pigs		Pigs, Sheep					
Problems of maternal behaviour & young	Mother preventing suckling	Pigs, Sheep, Horses		Pigs, Cattle, Sheep				Pigs, Cattle, Sheep, Chickens & Turkeys	
Problems of maternal behaviour & young	Mis-mothering (wrong young)			Pigs, Cattle, Sheep		Pigs, Cattle, Sheep, Horses			
Problems of maternal behaviour & young	Non-recognition of mother by young								
Problems of maternal behaviour & young	Non-suckling of young	Pigs, Cattle, Sheep, Horses		Pigs, Cattle		Pigs, Cattle, Sheep, Horses			
Problems of maternal behaviour & young	Cannibalism of young	Pigs	Pigs	Pigs, Horses		Pigs	Pigs	Pigs	
Stereotypies	Intersuckling				Cattle	Cattle			
Stereotypies	Crib biting Wood eating	Horses	Pigs, Cattle, Horses		Cattle		Pigs, Cattle		
Stereotypies	Self-licking	Pigs, Cattle, Sheep, Horses	Pigs, Cattle		Cattle		Pigs, Cattle		
Stereotypies	Licking surrounds	Pigs, Cattle, Sheep	Pigs, Cattle		Cattle				
Stereotypies	Weaving	Pigs, Horses	Pigs, Horses						
Stereotypies	Feather picking	Chickens & Turkeys	Chickens & Turkeys						
	Abnormal defaecation behaviour			Pigs			Pigs	Pigs	
	Intake reduced	Pigs, Cattle, Horses, Chickens & Turkeys	Pigs, Cattle, Horses	Pigs, Cattle	Cattle			Pigs, Cattle, Sheep, Chickens & Turkeys	
	Intake increased		Pigs, Cattle, Horses						Cattle, Sheep, Horses, Chickens & Turkeys

Key: Pigs — Cattle — Sheep — Horses — Chickens & Turkeys

Poor ventilation	Draughts	High lighting intensity	Low lighting intensity	Monotonous environment	Monotonous food	Powdered food	Isolation	Group too large	Absence of male

BIBLIOGRAPHY

ADER, R. & CONKLIN, P.M. 1963. Handling of pregnant rats: effects on emotionality of their offspring. Science 142, 411-412.

ADER, R. & FRIEDMAN, S.F. 1964. Psychological factors and susceptibility to disease in animals. In Medical Aspects of Stress in a Military Climate 457-470.

ALBRIGHT, J.L., GORDON, W.P., BLACK, W.C., DIETRICH, J.P., SNYDER, W. W. & MEADOWS, E.C. 1966. Behavioural responses of cows to auditory conditioning. J. Dairy Sci. 4, 104-106.

ANDERSON, James 1797. Essays relating to agriculture and rural affairs. 4th ed. London.

ANDREW, R.J. 1957. The influence of hunger on aggressive behaviour in certain buntings of the genus Emberiza. Physiol. Zool. 30, 177-185.

ANDREW, R.J. & ROGERS, L.J. 1972. Testosterone, search behaviour and persistance. Nature Lond. 237, 343-346.

ANONYMOUS, 1968. Beef Recording Association Report, p.14.

ANONYMOUS, 1973. Milk Cost Tables (1972-1973) Milk Marketing Board.

ANTRUM, H. & VON HOLST, D. 1968. Sozialr, stress bei Tupajas (Tupaia glis), und sein Wirkunfanf Wachotum, Kerpergeurcht und Fortpflanzung. Z. Vergl. Physiol. 58, 347-355.

ARCHER, J. 1970a. Effects of population density on behaviour of rodents. In Social Behaviour in Birds and Mammals. Ed. J.H. Crook, 169-210.

ARCHER, J. 1970b. The effect of strange male odour on aggressive behaviour in male mice. J. Mammal 19, 572-575.

ARCHER, J. 1970c. Psychological stress and adrenocortical function. (unpub. mss).

ARCHER, J. 1973. Effects of testosterone on immobility responses in young male chicks. Behav. Biol. 8, 551-556.

ARCHER, J. 1973. A further analysis of responses to novel environment by testosterone-treated chicks. Behav. Biol. 9, 389-396.

ARCHER, J. 1974. The organisation of aggression and fear in vertebrates. In Perspectives in Ethology. 2nd Ed. Ed. P. Bateson and P. Klopfer.

ARNOLD, G.W. 1962 Factors within plant associations affecting the behaviour and performance of grazing animals. In Grazing in Terrestrial and Marine Environments. Ed. A.J. Crisp. Symp. Brit. Ecol. Soc. 4. Oxford Blackwell.

ARTHUR, G.H. 1961. Some observations on the behaviour of parturient farm animals with particular reference to cattle. Vet. Review.

ARTHUR, G.H. 1965. A rational veterinary approach to equine parturition. Brit. Equine Vet. Assoc. 4th Cong.

AZRIN, N.H. 1964. Aggressive responses of paired animals. In Medical Aspects of Stress in a Military Climate, 329-345.

BACA, A.S.F., GONZALES, G.E., MADARIEGUE, M.F. & NOLTE, M.M. 1965. Anim. Br. Abstr. 1965. 33, 2054.

BAENNINGER, L.P. 1966. The reliability of dominance order in rats. Anim. Behav. 14, 367-371.

BAENNINGER, L.P. 1973. Aggression between weanling Peromyscus and Microtus. Anim. Behav. 21, 335-337.

BAINBRIDGE, J.G. 1969. Effect of chlorpromazine on social rank in rats. Br. Vet. J. 125, 125-252.

BAINBRIDGE, J.G. 1970. The effect of certain drugs on appetite. Brit. Vet. J. 126, 660.

BALDWIN, B.A. 1969. The study of behaviour in pigs. Brit. Vet. J. 125, 281.

BALDWIN, B.A. & INGRAM, D.L. 1967. Behavioural thermo-regulation in pigs,. Physiol. & Behav. 2, 15-21.

BALDWIN, B.A. & SHILLITO, E.E. 1974. The effects of ablation of the olfactory bulbs on parturition and maternal behaviour in Soay sheep. Anom. Behav. 22, 220-223.

BANE, A. 1963. Experimental studies on serving behaviour in dairy bulls. Int. Vet. Cong. 53, 709-712.

BANERJEE, U. 1971. An enquiry into the genesis of aggression in mice induced by isolation. Behav. XL, 86-99.

BAREHAM, J.R. 1971. Imprinting and neonatal behaviour. Society Vet. Ethology, Bristol, September 1971.

BARKER, C.A.V. 1960. Discussion of service behaviour of bulls. Ann. Meet. Ontario Assoc. Artificial Breeders, Toronto.

BARON, A. & KISH, G.B. 1962. Early social isolation as a determinant of aggressive behaviour in the domestic chick. J. Comp. Physiol. Psychol. 53, 459-463.

BARYSHNIKOV, I.A. & KOKORINA, E.P. 1959. R. Vet. Int. Dairy Conf. 15, 46.

BARYSHNIKOV, I.A. & KOKORINA, E.P. 1964. Higher nervous activity of cattle. Dairy Sci. 26, 97.

BASKIN & GAUTHIER-PILTERS. In The Behaviour of Ungulates and its Relations to Management. Ed. V. Geist and F. Walther. Published by IUCN. Morges.

BAYER, C. 1929. Beitrage zuis zweikamp. Zeits Phycol. 112, 1-54.

BEACH, F.A. 1942. Comparison of copulatory behaviour of male rats raised in isolation, cohabitation and segregation. J. Genet. Psychol. 60, 121-136.

BEAMER, W., BERMANT, G. & CLEGG, M.T. 1969. Copulatory behaviour of the ram (Ovis aries). 11. Factors affecting copulatory satiation. Anim. Behav. 17, 706-711.

BEARSE, G.E., BERG, L.R., McCLARY, C.F. & MILLER, V.L. 1949. The effect of pelleting chickens' rations on the incidence of cannibalism. Poul. Sci. 28, 756.

BEEMAN, E.A. & ALLEE, W.C. 1945. Some effects of thiamin on the winning of social contacts in mice. Physiol. Zool. 18, 195-221.

BEILHARZ, R.G. 1968. Effect of stimuli associated with the male on litter size in mice. Austr. J. Biol. Sci. 21, 583-585.

BEILHARZ, R.G. & COX, D.F. 1967. Social dominance in swine. Anim. Behav. 15, 117-122.

BEILHARZ, R.G. & MYLREA, P.J. 1963. Social position and behaviour of dairy heifers in yards. Anim. Behav. 11, 522-528.

BELL, R.H.V. 1970. The use of the herb layer by grazing ungulates in the Serengeti. In Animal Populations in Relation to their Food Resources. Ed. A. Watson, 111-125. Oxford. Blackwell.

BELLINGER, L.L. & MENDEL, V.E. 1974. A note on the reproductive activity of Hampshire and Suffolk ewes outside the breeding season. Anim. Prod. 19, 123-126.

BENNETT, E.L.M., DIAMOND, C., KRECH, D. & ROSENZWEIG, M.R. 1964. Chemical and anatomical plasticity of the brain. Science 146, 610-618.

BERKSON, G. 1967. Abnormal stereotyped motor acts. In Comparative Psychopathology, Animal and Human. Ed. J. Zubin and H.F. Hunt. Grove & Tratton, 76-94.

BERKSON, G. & DAVENPORT, R.K. 1962. Stereotyped movements of mental defectives. (1) Initial survey. Amer. J. Ment. Defic. 66, 849-852.

BERKSON, G., MASON, W.A. & SAXON, S.V. 1963. Situation and stimulus effects on stereotyped behaviour of chimpanzees. J. Com. Physiol. Psychol. 56, 786-792.

BERKSON, G. & MASON, W.A. 1964. Steroetyped behaviour of chimps in relation to general arousal and alternative activities. Perceptual and Motor Skills 19, 635-652.

BERKOWITZ, L. 1962. Aggression. McGraw Hill, N.Y.

BERNSTEIN, I.S. 1970. Primate status hierarchies. In Primate Behaviour. Ed. L.A. Rosenblum. 1. 71-109.

BERNSTEIN, L. 1957. The effects of variations in handling upon learning and retention. J. Comp. Physiol. Psychol. 50, 162-167.

BIRKE, L.I.A. 1974. Social facilitation in the Bengalese finch. Behav. 48, 111-122.

BIRKE, L.I.A. & CLAYTON, D.A. (in press) . Social Facilitation: a Review.

BISHOP, M.W.H. & WALTON, A. 1960. Spermatogenesis and the structure of mammalian spermatozoa. In Marshall's Physiology of Reproduction, Vol 1. Ed. Parkes. Longmans, Green & Co.

BLAUVELT, M. 1954. Mother-newborn relationship in goats. Group. Proc. 1st Conf. Josian Macy Jr Found. 221.

BLAXTER, K.L. 1962. The energy metabolism of ruminants. Chicago. Charles Thomas.

BLAXTER, K.L. 1967. The energy metabolism of ruminants, Hutchinson. London.

BLAXTER, K.L. 1974. Energy in agriculture. Nature, Lond. 252, 531.

BLAXTER, K.L. & FRENCH, T.W. 1944. Experiments on the use of home-grown foods for milk production. J. Agric. Sci. 34, 212-222

BLURTON-JONES, N.G. 1968. Observations and experiments on the causation of threat displays of the Great Tit (Parvs major). Anim. Behav. Monograph 1, 75-158.

BOSKA, S.C. , WEISMAN, H.M. & THAR, D.H. 1966. A technique for inducing aggression in rats utilizing morphine withdrawal. Psychol. Rec. 16, 541-543.

BOUISSOU, M.F. 1970. Technique de mise en evidence des relations hierarchiques dans une groupe de bovins domestique. Rev. Comp. Anim. 4, 66-69.

BOUISSOU, M.F. 1971. Effet de l'absence d'informations optiques et de contact physique sur la manifestation des relations hierarchiques chez les bovins domestiques. Ann. Biol. Anim. Bioch. Biophys. 11, 191-198.

BOUISSOU, M.F. & SIGNORET, J.P. 1970. La hierarchie sociale chez les mammiferes. Rev. Comp. Animal. 2, 43-61.

BRADY, J.V. 1964. Experimental studies of psycho-physiological responses to stressful situations. In Medical Aspects of Stress in a Military Climate, 271-289.

BRAMBELL, F.W.K. 1965. Report of technical committee on welfare of animals under intensive husbandry system. H.M.S.O. London.

BRANTAS, G.C. 1968a. On the dominance order in Friesian-Dutch dairy cows. Zeits. fur Tierzuchtung and Zuchtungsbiol. 84, 127.

BRANTAS, G.C. 1968b. Training, eliminative behaviour and resting behaviour of Friesian-Dutch cows in the cafetaria stable. Zeits. fur Tierzuchtung und Zuchtungsbiol. 85, 64-77.

BROWN, V.M. Aggressive behaviour in the cod. (Yadus callarias C.). Behav. 18, 107-197.

BRUNSDON, R.V. 1965. Internal parasites and sheep production. Ruaraka Farmers' Conf., New Zealand. 43-60.

BRYANT, M.J. 1970. The influence of population density and group upon the behaviour of the growing pig. Ph. D. Thesis. University of Liverpool.

BRYANT, M.J. 1971. The social behaviour of the pig. Proc. 2nd British Council Course on Management and Diseases of Pigs. September. 12-25.

BUECHNER, H.K. 1961. Territorial behaviour in the Uganda kob. Science 133, 698-699.

CALHOUN, J.B. 1962. Behavioural Sink. In Roots of Behaviour. Ed. E.L. Bliss. Harper and Row. New York.

CALHOUN, J.B. 1963. The social use of space. In Viewpoint in Biology, Vol, 3. Ed. W.V. Mayer and R.G. van Gelder. Butterworth & Co., London.

CAMPBELL, Q.P. & BOSMANS, S.W. 1964. The effect of time and method of castration and docking on growth and carcass quality of Dorper lambs. Proc. S.A. Soc. Anim. Prod. 136-138.

CAMPLING, R.C. & FREER, M. 1966. Factors affecting the voluntary intake of food by cows. 8. Experiments with ground pelleted roughages. Br. J. Nutr. 20, 229-243.

CANDLAND, D.K. & BLOOMQUIST, D.W. 1965. Interspecies comparisons of the reliability of dominance orders. J. Comp. Physiol. Psychol. 59.

120   Bibliography

CASTLE, M.E., FOOT, A.S. & HALLEY, R.J. 1950. Some observations of the behaviour of dairy cattle with particular reference to grazing. J. Dairy Res. 17, 215-229.
CHAMPAYNE, J.R., CARPENTER, J.W., HENTGES, J.F., PALMER, A.Z. & COGER, M. 1969. Feedlot performance and carcass characteristics of young bulls and steers castrated at 4 ages. J. Anim. Sci. 29, 887-890.
CHANCE, M.R.A. 1954. The suppression of audio-genic hyper-excitement by learning in Peromyscys maniculatus. Brit. J. Anim. Behav. 11, 31-35.
CHANCE, M.R.A. & MACKINTOSH, J.H. 1962. The effects of caging. Coll. papers of Lab. Animal Centre 11, 59-64.
CHMIELNICK, H. 1965. Anim. Breed. Abstr. 302, 4, 518.
CHRISTIAN, J.J. 1963. Endocrine aductive mechanisms and the physiological regulation of population growth. In Mammalian Polulations. Ed. W.V. Meyer and R.G. Van Gelder. Academic Press New York.
CHRISTIAN, J.J. & DAVIS, D.E. 1964. Endocrines, behaviour and population. Science 146, 155-160.
CLARK, J.N. 1965. Methods of lamb castration. Proc. of Ruaraka Farmers' Conf. New Zealand.
CLARK, J.R. 1956. The aggressive behaviour of the vole. Behav. 9, 1-23.
CLARK, L.D. 1964. Aggressive behaviour and factors affecting it. In Medical Aspects of Stress in a Military Climate 311-328.
CLARKE, K.W. 1963. Stocking rate and sheep/cattle interactions. Wool Tech. & Sheep Breed. 10, 27-32.
CLAWSON, A.R. & BARRICK, E.R. 1962. The influence of number of pigs per pen or floorspace allotment on performance of growing pigs. N.C. State Coll. Publ. 82.
COFFEY, D.J. 1971. Some effects of social isolation. Brit. V. Jour. 129.
COFFEY, D.J. 1971. The concept of stress. Brit. V. Jour. 130, 91.
COLEMAN, J.H. 1950. Teaser ram induces earlier mating. Agric. Graz. NSW 61, 440.
COLLIAS, N.E. 1950. Some variations in grouping and dominance patterns among birds and mammals. Zoologica 35, 97-119.
COLLIAS, N.E. 1956. The analysis of socialisation in sheep and goats. Ecology 37, 228-239.
COOK, W.R. 1971. The cause of ear, nose and throat diseases in the horse. British Racehorse, August.
COOPER, J.J. & LEVINE, R.C. 1973. Effects of social interaction on eating and drinking on two sub-species of deer mice. Anim. Behav. 21, 421-428.
CRESWELL, E., ASH, R.W., BOYNE, A.W. & GILL, J.C. 1964. Growth and carcass characteristics of entire crossbred lambs compared with lambs 'partially' or 'fully' castrated. Vet. Rec. 76, 1472-74.
CREW, F.A.E. & MIRSKAIA, L. 1931. The effects of density on an adult mouse population. Biologica Generalis 7, 239-250.
CROSS, B.A. 1955. Neuro-hormonal mechanisms in emotional inhibition of milk ejection. J. Endocrin. 12, 29.
CROWFORD, M.P. 1939. The social psychology of the vertebrates. Psychol. Bull. 36, 407-446.
CROWLEY, J.P. & DARBY, T.E. 1970. Observations on the fostering of calves for multiple suckling systems. Br. Vet. J. 126, 658.
CULLISON, A.E. 1961. Effect of physical form of the ration on steer performance and certain rumen phenomena. J. Agric. Aci. 20, 478-483.
CULPIN, S., EVANS, W.M.R. & FRANCIS, A.C. 1964. An experiment on mixed pasture. Exper. Husb. 10, 29-38.

DALE, J. 1965. Grazing experiments in Snowdonia. In Grazing Experiments and the Use of Grazing as a Conservation Tool. Monks Wood Experimental Station, England. Symp. No. 2.
DAVIS, C.R., AUTREY, K.M., HERLICH, H. & HAWKINS, G.E. 1967. Outdoor

individual portable pens compared with conventional housing for raising dairy calves. J. Dairy Sci. 562-570.

DAWSON, F.L.M. 1970. Some aspects of the behaviour of stallions and mares. Soc. Vet. Ethol. Brit. Vet. J. 126, 659.

D'ALBA, J. 1960. Milking with calf and reproductive efficiency in the cow. Animal Breed. Absts. 30, 113.

DEAS, D.W. 1970. New light on infertility. Dairy Farmer Suppl.

DENENBERG, V.H. 1964. Some relationships between strong ('stressful') stimulation in infancy and adult performance. In Medical Aspects of Stress in a Military Climate 297-310.

DENENBERG, V.H. & WHIMBEY, A.E. 1963. Infantile stimulation in animal husbandry. J. Comp. Physiol. Psychol. 56, 877-878.

DICKSON, D.P., BARR, G.R. & WIECKEIR, D.A. 1967. Social relationship of dairy cows in a feed lot. Behav. 29, 195-203.

DIETRICH, J.P., SNYDER, W.W., MEADOWS, C.E. & ALBRIGHT, J.L. 1965. Rank order in dairy cows. American Zoologist 5, 713 (Abstr.).

DIMITROV, D.G. and NEICEV, O. 1964. The effects of various methods of castration on the wool production of ram lambs. Anim. Breed. Abstr. 34, 63.

DODD, F.H., FOOT, A.S., HENDRICHS, E. & NEAVE, F.K. 1950. Effect of subjecting dairy cows for a complete lactation to a rigid control of duration of milking. J. Dairy Res. 17, 107-116.

DONALDSON, S.L. 1967. The effect of early feeding and rearing experience on dominance, aggression and submission behaviour in young heifer calves. MS Thesis, Purdue University.

DONALDSON, S.L. & JAMES, J. 1964. Connection between crush order and pregnancy in cows. Anim. Behav. 11, 286.

DONALDSON, S.L., ALBRIGHT, J.L. & BLACK, W.C. 1972. Primary social relationships and cattle behaviour. Proc. Indiana Ac. Sci. 1971, Vol. 81.

DONALDSON, S.L., BLACK, W.C. & ALBRIGHT, J.L. 1966. Effects of early feeding in rearing experience on dominance, aggressive and submissive behaviour in young heifer calves. Amer. Zoologist 6, 247.

DONEY, J.H., GUNN, R.G. & GRIFFITHS, J.G. 1973. The effects of premating 'stress' on the onset of oestrus and on ovulation rate in the Scottish black-faced ewes. J. Reprod. Fert. 35, 381-384.

DORAN, C.W. 1943. Activities and grazing habits of sheep on summer ranges. J. Forestry 41, 253-258.

DOVE, W.F. 1935. American Naturalist, 19, 469.

DRAPER, W.A. 1965. Sensory stimulation and Rhesus monkey activity. Perceptual and Motor Skills 21, 319-322.

DREWS, D.R. 1973. Group formation in captive Galago crassicaudatus. Notes on the dominance concept. 1. Zeit. fur Tierpsychol. 32, 425-435.

DUN, R.B. 1963. The influence of the pole gene and/or castration on production characteristics of male Merino sheep. Aust. J. Expt. Agric. Anim. Hus. 3, 262-265.

DUNCAN, I.J.H., HORNE, A.R., HUGHES, B.O. & WOOD-GUSH, D.G.M. 1970. The pattern of food intake in female brown leghorn fowls as recorded in a skinner box. Anim. Behav. 18, 245-255.

DUNCAN, I.J.H. & HUGHES, B.O. 1972. Free and operant feeding in domestic fowls. Anim. Behav. 20, 775-777.

DUNCAN, I.J.H. & WOOD-GUSH, D.G.M. 1971. Frustration and aggression in the domestic fowl. Anim. Behav. 19, 500-504.

DUNCAN, I.J.H. & WOOD-GUSH, D.G.M. 1972. Thwarting of feeding behaviour in the domestic fowl. Anim. Behav. 20, 44.

DUNCAN, I.J.H. & WOOD-GUSH, D.G.M. 1974. The effects of Rauwalfia tranquillizer on stereotyped movements in frustrated domestic fowl. App. Anim. Ethol. 1, 67-76.

DUTT, R.H., SIMPSON, E.C., CHRISTIAN, J.C. & BARNHART, C.E. 1959. Identification of preputial glands as the site of production of sexual odour in the boar. J. Anim. Sci. 18, 1557.

EDGAR, D.G. 1965. Talking about tupping. Ruaraka Farmers' Conf. New Zealand, 61-69.

EISENBERG, J.S. 1967. The social organizations of mammals. Kukenthal Handbook. Zool. 10, 7.

EKESBO, I. 1974. Swedish studies on the impact of the environment on farm animals. B. Vet. J. 130, 92.

ELLIOT, O. & KING, J.A. 1960. Effect of early food deprivation upon later consummatory behaviour in puppies. Psychol. Rep. 6, 391-400.

ELOFF, H.B., REINECKE, J. & SKINNER, J.E. 1965. Influence of method of and age at castration on growth and development of oxen under ranching conditions, S.A. J. Agric. Sci. 8, 239-252.

ESPMARK, Y. 1971. Individual recognition by voice in reindeer mother-young relationship. Field observations and play back experiments. Behav. XI, 295-301.

ESSLEMONT, R.J. & BRYANT, M.J. 1974. Manifestation and direction of oestrus in a large dairy herd. Br. Vet. J. 129, 504.

ESTES, R.D. 1966. Behaviour and life history of the wildebeest, Connochaetes Taurinus. Burchell. Nature 212, 999-1000.

EWBANK, R. 1962. Crib biting in cattle. J.S.A. Vet. Med. Ass. 36, 1965.

EWBANK, R. 1969. Effect of stocking rates and group size on the social behaviour of the pig. Proc. 1st Int. Pig Vet. Soc. Conf. Cambridge, 61.

EWBANK, R. 1973. Activities of pigs in enclosed and outdoor environments. ASAB. Meeting December.

EWBANK, R. 1973. Preliminary observations on apparent dominance and leadership hierarchies in fattening sheep. Br. Vet. J. 129, 501.

EWBANK, R. & BRYANT, M.J. 1969. Some aspects of high stocking rates upon the behaviour of pigs. Proc. Soc. Vet. Ecol. Brit. Vet. J. 125, 250.

EWBANK, R. & BRYANT, M.J. 1969. The effects of population density upon the behaviour and economic performance of fattening pigs. Farm Blds. Progress 18, 14-15.

EWBANK, R. & BRYANT, M.J. 1972. Some effects of stocking rate upon agonistic behaviour in groups of growing pigs. Anim. Behav. 20, 21-28.

EWBANK, R. & MEESE, G.B. 1971. Aggressive behaviour in groups of domesticated pigs on removal and return of individuals. (In press).

EYLES, D.E. 1959. Feeding and management of pregnant sows on pasture. Anim. Prod. 1, 41-50.

FAICHNEY, C.J. 1968. The effect of frequency of feeding on the ulilization of roughage diets by sheep. Austr. J. Agric. Res. 19, 813-819.

FARRIS, E.J. 1954. Activity of dairy cows during oestrus. J. Vet. Med. Soc. 125, 117-120.

FENTON, P.K., ELLIOT, J. & CAMPBELL, R.C. 1969. The effects of different methods of castration on the growth rate and well-being of calves. Vet. Rec. 70, 101-106.

FERGUSON, W. 1968. Abnormal behaviour in domestic birds. In Abnormal Behaviour of Animals. Ed. M.W. Fox. W.B. Saunders.

FERGUSON, W. 1969. The role of social stress in epidemiology. Brit. Vet. J. 125, 253-254.

FERGUSON, W., HERBERT, W.J. & McNEILLAGE, D.J.C. 1970. Infectivity and virulence of Tripnosoma (tripnosoon) brucii in mice. Trop. Anim. Health and Prod. 2, 59-64.

FISH, F.J. 1962. Schizophrenia. John Wright & Sons.

FLETCHER, K. & LINDSAY, D.R. 1968. Sensory involvement in the mating behaviour of domestic sheep. Anim. Behav. 16, 410-414.

FLETCHER, T.J. & SHORT, R.V. 1974. Restoration of libido in castrated red deer (Cervus elaphus) with oestradiol-17B. Nature 248, 616-618.

FOOT, T. & DONEY, J.M. 1971. Personal communication. Hill Farming Research

Organization, Edinburgh.

FORBES, J.J., RAISIN, A.M., URWIN, J.D. & ROBINSON, K.L. 1968. On comparison of steers and partial castrates for beef production, 1. The performance of animals on a concentrate diet. Rec. Agric. Res. 16, 127-132.

FOX, M.W. 1965a. Environmental factors influencing stereotyped allelomimetic behaviour in animals. Lab. Anim. Care 15, 363-370.

FOX, M.W. 1965b. Canine behaviour. Charles C. Thomas, Illinois.

FOX, M.W. 1968. Socialisation, environmental factors and abnormal behavioural development in animals. In Abnormal Behaviour of Animals. Ed. M.W. Fox. W.B. Saunders.

FRADRICH, H. 1974. A comparison of behaviour in the Suidae. In The Behaviour of Ungulates in Relation to Management. Ed. V. Geist and F. Walter. 133-144.

FRASER, A.F. 1959. Influence of psychological and other factors on reaction time in the bull. Cornell Vet. L. 1960.

FRASER, A.F. 1961. Reproduction in goats. M.Sc. Thesis, University of Toronto.

FRASER, A.F. 1968a. The ram effect in breeding results in Suffolk ewes. Scot. Agric. August.

FRASER, A.F. 1968b. Reproductive behaviour in ungulates. Acad. Press.

FRASER, D. 1974. The vocalisations and other behaviour of growing pigs in an 'open field' test. Appl. Anim. Ecol. 1, 3-16.

FREDEEN, H.T. & JONSSON, P. 1957. Genetic variance and co-variance in Danish Landrace swine as evaluated under a system of individual feeding of progeny tested groups. Z. Tierz. Zuct. Biol. 70, 348-363.

FREDERICSON, E. 1950. The effects of food deprivation on competitive and spontaneous combat in C57 black mice. J. Psychol. 29, 89-100.

FREEMAN, D.M. 1971. Stress and the domestic fowl: a physiological appraisal. World Poultry Sci. J. 27, 263-275.

FRIEDBERGER & FROHNER 1904. Veterinary Pathology Vol. 1. The Chappel River Press, Kingston.

GADD, J. 1967. Tail biting. National Hog Farmer 12, 24.

GARTLAN, J.S. 1968. Structure and function in primate society. Folia primat. 8, 89-120.

GASSNER, F.X., REIFENSTEINER, E.C., ALGES, J.W. & MATTOX, W.E. 1958. Rec. Prog. Hormone Res. 14, 183. New York Academic Press.

GEIST, V. 1971. Mountain sheep. A study in behaviour and evolution. Univ. Chic. Press.

GERLBACH, G.D., BECKER, D.E., COX, J.L., HARMON, B.G. & JENSEN, A.U. 1966. Effects of floor space allowance and number per group on performance of growing finishing swine. J. Anim. Sci. 25, 386-391.

GINSBERG, B. & ALLEE, W.C. 1942. Some effects of conditioning on social dominance and subordination in inbred strains of mice. Physiol. Zool. 4, 485-506.

GORDON, J.G. & TRIBE, D.E. 1951. The self selection of diet by pregnant ewes. J. Agric. Sci. 41, 187-190.

GRANT, E.C. & CHANCE, M.R.A. 1958. Rank order in caged rats. Anim Behav. VI, 183-194.

GRAY, J.A. 1972. The psychology of fear and stress. Weidenfeld & Nicholson, London.

GREY, J.A. 1971. Acta Psychol. 35, 29-46.

GROSS, W.R. & SIEGEL, H.S. 1965. The effect of social stress on resistance to infection with Escherichia coli or Mycoplasma gallisepticum. Poultry Sci. 44, 998-1001.

GRUBB, P. 1974. Social organization of Soay sheep and the behaviour of ewes and lambs. In Island Survivors. Ed. P.A. Jewell, C. Milner and J. Morton Boyd. Athlone Press.

GRUBB, P. & JEWELL, P.A. 1966. Social grouping and home range in feral Soay sheep. In Play, Exploration and Territory in Mammals. Ed. P.A.

Jewell and C. Loisos. Academic Press.
GUHL, A.M. 1953. Social behaviour of the domestic fowl. Kansas State Coll. Agric. Expt. St. Tech. Bull. 73.
GUHL, A.M. & ALLEE, W.C. 1944. Some measurable effects of social organization in flocks of hens. Physiol. Zool. 320-347.
GUHL, A.M. & ATKESON, F.W. 1959. Social organization in a herd of dairy cows. Trans. Kansas Acad. Sci. 62, 80-87.
GUHL, A.M., COLLIAS, N.E. & ALLEE, W.C. 1945. Mating behaviour in the social hierarchy in small flocks of white leghorns. Physiol. Zool. 18, 365-390.
GUNDLACK, H. 1968. Brutfursorge, Brutpflege, Verhaltensont ogen und Tagesperiodik beim Europaischen Wildschwein (Sus scofa). Z. fur Tierpsychol. 25, 955-995.

HAFEZ, E.S.E. 1951. Mating behaviour in sheep. Nature 167, 777-778.
HAFEZ, E.S.E. 1968. Environmental effects of animal production. In Adaptation in Animals. Ed. E.S.E. Hafez.
HAFEZ, E.S.E. & SIGNORET, J.P. 1969. The behaviour of swine. In The Behaviour of Domestic Animals. Ed. E.S.E. Hafez. Baillière, Tindall.
HALE, E.S. & ALMQUIST, J.O. 1956. Effects of changes in the stimulus field on responsiveness of bulls to a constant stimulus animal. Anat. Rec. (Abstr.) 125, 607.
HAMILTON, D. & BATH, J.G. 1970. Performance of sheep and cattle grazed separately and together. Expt. Agric. and Anim. Husb. 10, 19-26.
HANCOCK, J. 1954. Studies of grazing behaviour in relation to grassland management. 1. Variations in grazing habits of dairy cattle. J. Agric. Sci. 44, 420-433.
HANSEN, T.E. & MANGOLD, D.W. 1960. Functional and basic requirements of swine housing. Agric. Engin. 41, 585-590.
HANSON, R.P. & LARS, K. 1959. Feral swine in South Eastern United States. J. Wildl. Man. 23, 64-74.
HARDISON, W.A., REID, J.T., MARTIN, C.M. & WOOLFOLK, P.G. 1954. Degree of herbage selection by grazing cattle. J. Dairy Sci. 37, 87-102.
HARLOW, H.F. 1932. Social facilitation of feeding in the rat. J. Genetic Psychol. 4l.
HARLOW, H.F., HARLOW, M.K. & HANSEN, E.W. 1963. The maternal affectional system of Rhesus monkeys. In Maternal Behaviour in Mammals. Ed. H.C. Rheingold, 254-281.
HART, G.C., MEAD, S.W. & REGAN, W.M. 1946. Stimulating the sex drive of bovine males in AI. Endocrin. 39, 221-223.
HASTIE, H. 1969. Observations in a barley beef unit. Brit. Vet. J. 125, 251.
HATCH, A.M., WIBERG, G.S., ZAWIDZKA, M. CANN., AIRTH, J.M. & GRICE, H.C. 1965. Isolation syndrome in the rat. Toxicology & Appl. Pharm. 7, 737-745.
HAWKER, R.W. 1963. Hypothalamic control of pituitary function. In Modern Trends in Human Reproductive Physiology. 1. Ed. Carey, Butterworths.
HAYS, R.C. & CARLEVARO, C.H. 1959. Introduction of oestrus by electrical stimulation. Am. J. Physiol. 196, 899-900.
HEDIGER, H. 1955. Studies of the psychology of animals in zoos and circuses. Butterworth Scientific Publ.
HEITMAN, H., HAHN, LE ROY, KELLY, C.F. & BOND, T.E. 1961. Space allotment and performance of growing finishing swine raised in confinement. J. Anim. Sci. 20, 543-546.
HERSHER, L., RICHMOND, J.B. & MOORE, A.U. 1963. Maternal behaviour in sheep and goats. In Maternal Behaviour of Mammals. Ed. H.L. Rheingold Wiley, N.Y. 203-232.
HOPPMEYER, L. 1969. Feather pecking in pheasants—an ethological approach to the problem. Dan. Rev. Game Biol. 6, 1-36.

125   Bibliography

HORRIDGE, P.A.S. 1970. The development of copulatory and fighting
behaviour in the domestic chick. D. Phil. Thesis, University of Sussex.
HOUSEMAN, R.A. 1973. A comparison of voluntary food intakes of boars, gilts
and castrates. 56th Meet. Soc. Anim. Prod. 2, 90.
HOYENGA, K.T. & AESHELMAN, S. 1969. Social facilitation of eating in the
rat. Psychon. Sci. 14, 129-141.
HUGHES, B.O. & DEWAR, W.A. 1971. A specific appetite for zinc in zinc
depleted domestic fowl. Br. Poult. Sci. 12, 255-258.
HUGHES, B.O. & DUNCAN, I.J.U. 1972. The influence of strain and
environmental factors upon feather pecking and cannibalism in fowls. Br.
Poult. Sci. 13, 525-547.
HUGHES, B.O. & WOOD-GUSH, D.G.H. 1971. Investigations into specific
appetites for sodium and thiamine in domestic fowl. Physiol. and Behav. 6,
331-339
HUNTER, R.F. 1960. Conservation and grazing. Inst. Biol. 7, 22-26.
HUNTER, R.F. 1964. Home range behaviour in hill sheep. In Grazing in
Terrestrial and Marine Environments 155-171. Dorking, England. Blackwell
Sci. Pub.
HUNTER, R.F. & MILNER, C. 1963. The behaviour of individual related and
groups of south country Cheviot sheep. Anim. Behav. 2 (4), 507-613.
HUNTER, W.K. & COUTTIE, M.A. 1969. The behaviour of groups of bulls at
lay-off. Proc. Soc. Vet. Ethol. in Brit. Vet. J. 125, 252.
HUTT, C. & HUTT, S.J. 1965. The effects of environmental complexity on
stereotyped behaviour of children. Anim. Behav. I3, 1-4.

JAMES, J.W. & FOENANDER, F. 1961. Social behaviour in domestic animals. I.
Hens in laying cages. Austr. J. Agric. Res. I2, 1239-1252.
JAMES, W.T. 1949. Dominant and submissive behaviour in puppies as indicated
by food intake. J. Genet. Psychol. 75, 33-43.
JANSEN, P.A.J., JAGENEAU, A.H. & NIENEGEERS, G.T.E. 1960. Effects of
various drugs on isolation-induced fighting behaviour of male mice. J.
Pharmacol. Expt. Therap. 129, 471-475.
JAY, P. 1965. The common langur of North India. In Primate Behaviour.
Ed. I. de Vore, Holt, Rinehart and Winston. N.Y. 197-249.
JERICHO, K.W.F. & CHURCH, T.L. 1972. Cannibalism in pigs. Can. Vet. Journ.
13, 156-159.
JEWELL, P., MILNER, C. & MARTIN BOYD, J. 1974. Island Survivors, The
Ecology of the Soay Sheep of St Kilda. Athlone Press.
JOHNSON, C.C. & WILKINSON, J.H. 1974. An investigation of the anabolic
effects of androstenedione in young ruminants. Anim. Prod. 18, 179-189.
JOHNSTONE-WALLACE, D.B. 1937. The influence of management and plant
associations on chemical composition of pasture plants. J. Amer. Soc.
Agron. 29, 441.
JOHNSTONE-WALLACE, D.B. & KENNEDY, K. 1944. Grazing management
practices and their relationship to the behaviour and grazing habits of
cattle. J. Agric. Sci. 34,190-197.
JOLLY, A. 1967. The Lemurs of Madagascar. Univ. Chicago Press.
JOUBERT, S.C. 1974. The social organisation of the roan antelope (Hippotragus
Equinus) and its influence on the spatial distribution of herds in the Kruger
National Park. In The Behaviour of Ungulates and its Relation to
Management. Ed. V. Geist and F. Walther. 661-676.

KARE, M.R., POND, W.C. & CAMPBELL, J. 1965. Observations on the taste
reactions in pigs. Anim. Behav. 13, 265-269.
KARE, M.R. & SCOTT, H.L. 1962. Nutritional value and feed acceptability.
Poultry Sci. 41, 276-278.
KECK, E., HALE, E.W., SCHEIN, M.W. & MILLER, R.C. 1962. A behavioural
inventory of sheep fed pelleted and baled alfalfa hay. J. Anim. Sci. 21, 138.
KEIPER, R.R. 1970. Studies of stereotype function in the canary (Serinus

canarius). Anim. Behav. 18, 353-357.

KERRUISH, B.M. 1955. The effect of sexual stimulation prior to service on behaviour and conception rate of bulls. Br. J. Anim. Behav. 3, 125.

KILEY, M. 1965. The behaviour of a small herd of feral Camargue ponies. (Unpub. manuscript).

KILEY, M. 1969. Some displays in ungulates, canids and felids with reference to causation. D. Phil. Thesis. Univ. Sussex.

KILEY, M. 1972. The vocalizations of ungulates. Zeit. fur Tierpsychol. 31, 171-222.

KILEY, M. 1974. The social organization of a captive group of eland. A.S.A.B. Conf. Cambs. July.

KILGOUR, R. 1969. Social behaviour in the dairy herd. N.Z. Dept. Agric. Ruaraka Pub. 459.

KILGOUR, R. & ALBRIGHT, J.L. 1971. Control of voiding habits. N.Z. J. Agric. 122, 31-33.

KILGOUR, R. & CAMPION, R.N. 1973. The behaviour of entire bulls of different different ages at pasture. Proc. N.Z. Soc. Anim. Prod. 33, 125-138.

KILGOUR, R. & DE LANGEN, H. 1970. Stress in sheep resulting from management practices. N.S. Dept. Agric. Ruaraka Agric. Res. Centre, N.Z.

KILGOUR, R. & SCOTT, T.H. 1959. Leadership in a herd of dairy cows. Proc. N.Z. Soc. Anim. Prod. 19, 36.

KLOPFER, F.D. 1961. Early experience and discrimination learning in swine. Am. Zool. I, 366. Abstr. 38.

KLOSTERMAN, E.W., KUNKLE, L.E., GERLAUGH, P.E. & CAHILL, V.R. 1954. The effect of age of castration upon rate and economy of gain and carcass quality of beef calves. J. Anim. Sci. 13, 817.

KOROTKOV, V.I. 1966. Is it more advantageous to fatten wethers or ram lambs? Anim. Breeding Ext. 391, 35.

KOSYH, A.P. 1962. Meat production is increased by partial castration. Anim. Breed. Abstr. 31, 358.

KRATZER, D.D. 1971. Learning in farm animals. J. Anim. Sci. 31, 1268-1273.

KRECH, D., MARK, R., ROSENZWEIG & BENNETT, E.L. 1960. Effects of environmental complexity and training on brain chemistry. J. Com. Physiol. Psychol. 6, 53, 509-519.

KRISTJANSSON, F.K. 1957. A note on the use of chloropromazine in the treatment of extreme nervousness and savageness in farrowing sows. Can. J. Comp. Med. & Vet. Sci. 21, 389-390.

KRUIJT, J.P. 1964. Ontogeny of social behaviour in Burmese Red Jungle Fowl (Gallus gallus spadiceus) Bonnaterre. Behav. Suppl. 12.

KUDRYARTZEV, A.A. 1962. Higher nervous activity and the physiology of the senses in lactating cows. Int. Dairy Conf. 160, 565-572.

KUO, Z.Y. 1960. Studies on the basic factors in animal fighting. II Nutritional factors affecting fighting behaviour in quails. J. Genet. Psychol. 96, 207-216.

LAING, R.D. & ESTERSON, A. 1964. Sanity, Madness and the Family, Tavistock.

LAL HARBANS & ROBINSON, M.A. 1965. Fixation of maladaptive stereotyped behaviour through operant conditioning. Amer. Soc. Zool. (abstr.) 6, 3, 295-296.

LAMPREY, H.F. 1963. Ecological separation of the larger mammal species in the Tarangire Game Reserve, Tanganyika. E. Afr. Wildl. J. 1, 63-92.

LANCASHIRE, J.A. & KEOGH, R.G. 1966. Some aspects of the behaviour of grazing sheep. Proc. N.Z. Anim. Prod. 26, 22-35.

LARSEN, H.J. 1963. J. Anim. Sci. 22, Abstr. 1134.

LEAVER, J.D. 1970. A comparison of grazing systems for dairy herd replacements. J. Agric. Sci. 265-272.

LEAVER, J.D. & YARROW, N.H. 1970. Cubicle housing for calves. NAAS Qrt. Rev. 86.

LEHRMAN, D.S. 1961. Hormonal regulation of parental behaviour in birds and infrahuman mammals. In Sex and Internal Secretions. Ed. W.C. Young, 3rd Ed. Baltimore, 1268-1382.

LEMMON, W.B. & PATTERSON, G.H. 1964. Depth perception in sheep. Effects of interrupting mother-neonate bond. Science 145, 835-836.

LENT, P. 1974. Mother-young relationships. In Ungulate Behaviour and its Relation to Management. Ed. V. Geist and F. Walther I.U.C.N. 1-24.

LEVINE, S. 1957. Infantile experience and resistance to physiological stress. Science 126, 405.

LEVINE, S. 1962. Emotionality and aggressive behaviour in the mouse as a function of infantile experience. Am. Psychol. 12, 410.

LEVY, D.M. 1944. On the problem of movement restraint. Amer. J. Orthopsychiat. 14, 644-671.

LIDDELL, H.S. 1956. Emotional hazards in animals and man. Springfield, Illinois.

LIDDELL, H.S. & ANDERSON, O.D. 1931. A comparative study of the conditioned motor reflex in the rabbit, sheep, goat and pig. Amer. J. Physiol. 97, 339.

LINCOLN, G.A., GUINESS, F. & SHORT, R.U. 1972. The way in which testosterone controls the social and sexual behaviour of the red deer stag (Cervus elephas). Horm. & Behav. 3, 375-396.

LINDSAY, D.R. 1965. Importance of olfactory stimuli in mating behaviour of the ram. Anim. Behav. 13, 75-78.

LINNEAUS, C. 1749. Amoenitates Academicae, Vol. II, 225. London.

LLOYD, J.A. & CHRISTIAN, J. 1967. Relationship of activity and aggression to density in 2 confined populations of house mouse (Mus musculus). J. Mammal. 48, 262-269.

LOGAN, W.A. 1965. Influence of cage versus floor density and dubbing on laying house performance. Poultry Sci. 44, 975-979.

MCBRIDE, G. 1959. The influence of social behaviour on experimental design in animal husbandry. Anim. Prod. 1, 81-84.

MCBRIDE, G. 1960. Poultry husbandry and the peck order. Br. Poul. Sci, 1, 65-68.

MCBRIDE, G. 1964. Social behaviour of domestic animals. II. The effect of peck order on poultry productivity. Anim. Prod. 6, 1-7.

MCBRIDE, G. 1968. Behavioural measurement of social stress. In Adaptation of Domestic Animals. Ed. E.S.E. Hafez.

MCBRIDE, G., JAMES, J.W. & HODGENS, N. 1964. Social behaviour of domestic animals. IV. Growing pigs. Anim. Prod. 6, 129-139.

MCCLINTOCK, M.K. 1971. Menstrual synchrony and suppression. Nature 229, 244-245.

MACKINTOSH, J.H. 1970. The territorial behaviour of the laboratory mouse. Univ. Sussex Seminar.

MACMILLAN, K.C. 1970. Return intervals to first insemination and conception rates to second insemination in New Zealand dairy cattle. N.Z. J. Agric. Res. 13, 771-777.

MCPHEE, C.P., MCBRIDE, G., & JAMES, J.W. 1964. Social behaviour of domestic animals. III. Steers in small yards. Anim. Prod. 6, 9-15.

MARCUSE, F.L. & MOORE, A.U. 1944. Tantrum behaviour in the pig. J. Comp. Physiol. Psychol. 38, 1.

MARLER, P. 1956. Studies of fighting in Chaffinches. 3. Proximity as a source of aggression. Anim. Behav. 4, 23-30.

MARQUISS, M. 1974. Measurement of 'aggression' and 'dominance' in captive red grouse. A.S.A.B. Cambs, 1974.

MARSDEN, H. 1966. On feather pecking and cannibalism in pheasant and partridge chicks particularly in relation to amino-acid arginine. Actas. Vet. Scand. 7, 272-287.

MARSDEN, H.M. & BRONSON, F.H. 1964. Oestrus synchrony in mice: alteration by exposure to male urine. Science 144, 1469.

MARSHALL'S Physiology of Reproduction. 1965. Ed. A.S. Parkes. Longmans.

128   Bibliography

MARX, M.H. 1952. Infantile deprivation and adult behaviour in the rat. Retention of increased rate of eating. J. Comp. Physiol. Psychol. 45, 43-49.
MASON, W.A. 1960. Effect of social restriction on monkeys. J. Comp. Physiol. Psychol. 53, 582-589.
MASON, W.A. 1965. The social development of monkeys and apes. In Primate Behaviour. Ed. I. De Vore, N.Y. Holt, Rienhart and Winston. 514-543.
MAYERS 1966. The effect of density on sociality and health in mammals. Proc. Ecol. Soc. Aust. 1, 60-64.
MECH, C.D. 1962. Ecology of the timber wolf (Canis impus). Ph. D. Thesis. Purdue Univ.
MEESE, G.B. 1971. The role of sensory systems in pig group reorganization. A preliminary report. Br. Vet. J. 129, 502.
MEYER, J.H., LOFGREEN, G.P. & HULL, J.L. 1957. Selective grazing by sheep and cattle. J. Anim. Sci. 16.
MEYER-HOLZAPFEL, M. 1968. Abnormal behaviour of zoo animals. In Abnormal Behaviour of Animals. Ed. M.W. Fox. W.B. Saunders.
MILLER, M.W. & BEARSE, G.E. 1938. The cannibalism preventing properties of oat hulls. Poult. Sci. 17, 466-471.
MILLER, W.C. 1950. Variations in male sexual behaviour. Proc. Brit. Soc. Anim. Prod. Meeting 16.
MILOVANOV, V.R., SMIRNOV-UGRUJOMOV, D.V. 1940. Problems of the rational use of stud animals in the light of Pavlov's teaching. Anim. Breed. Abstr. 1943. 11, 96.
MILTON, W.E. 1956. The palatability of herbage on under-developed grasslands in west-central Wales. Empire J. Exper. Agric. 21, 116-122.
MINISTRY OF AGRICULTURE, FISHERIES & FOOD 1965. Cow behaviour in loose housing. Tech. Rep. No. 13.
MINISTRY OF AGRICULTURE, FISHERIES & FOOD 1971, Order of recommendations for the welfare of livestock. 1-4.
MITCHELL & BROADBENT, P.J. 1973. The effect of level and method of feeding milk substitute and housing environment on the performance of calves. Anim. Prod. 17, 245-256.
MITIG, N. 1968. Stimulation of sexual activity in sheep. World Rev. Anim. Prod. IV. 19-20, 82-87.
MOCALOVSKII, A.N. 1963. New methods of sterilising rams and the effect on physiological and production characters. Anim. Breed. Extr. 492.
MORGAN, M. 1973. The effect of post weaning environment on rearing in the rat. Anim. Behav. 21, 429-442.
MORRIS, D. 1963. The response of animals to restricted environments. Symp. Zool. Soc. London. 19, 99-119.
MORRISON, S.R., HINTZ, H.F. & GIVENS, B.L. 1968. A note on the effect of exercise on behaviour and performance of confined swine. Anim. Prod. 10, 341-344..
MULLEN, P.A. 1964. Brit. Vet. J. 120, 518-523.
MURIE, A. 1944. The wolves of Mount McKinley, Washington U.S. Govt. Print. Office (U.S. Dept. Interior, Fauna, Series No. 5).
MYERS, K. 1966. The effects of density on sociality and health in mammals. Proc. Aust. J. Ecol. 1, 40-64.
MYLREA, P.J. & BEILHARZ, R.G. 1964. Manifestation and detection of oestrus in heifers. Anim. Behav. XII, 25.

NERSESJAN, S.S. 1959. The use of vasectomised bulls as biological stimulators in controlling infertility in cows. Anim. Br. Abstr. 1962. 30, No. 220.
NICHOLSON, A.M. 1974. Social interaction in a small group of beef cattle. Unpubl. B. Sc. Project, Univ. Sussex.
NISSEN, H.W. 1956. Individuality in the behaviour of chimpanzees. Amer. Anthrop. 58, 407-413.
NOLAND, P.R., SCOTT, K.W. & BARGUS, C.A. 1965. Effect of housing and exercise in pigs. Abst. Farm. Build. 7, 85.

NYHON, J.F. 1964. A comparison of results obtained with artificial insemination and natural mating in a pig-breeding herd. Vet. Rec. 76, 656-658.

OLSON, W.C. 1929. Genesis of Nervous Habits in Children. Ch. VI. Minneapolis University Press.

ØRSKOV, E.R., BENZIE, D. & KAY, R.N. 1970. The effects of feeding procedure on closure of the oesophageal groove in young sheep. Br. J. Nutr. 24, 785-795.

OWEN, J.B. 1969. The intensification of sheep production. Outlook on Agric. 6, 36-40.

PAYNE, A.P. & SWANSON, H.H. 1972. The effect of sex hormones on agonistic behaviour of golden hamsters. (Mesocricetus ancatus). Waterhouse. Physiol. Behav. 8687, 691.

PEARSON, M. 1970. Causation and development of behaviour in the guinea pig. D. Phil. Thesis, Univer. Sussex.

PEART, J.N. 1962. Increased production from hill pasture. Scottish Agriculture. Winter 1962-1963.

PEPELKO, W.E. & CLEGG, M.T. 1965. Studies of mating behaviour and some factors influencing the sexual response in male sheep. Ovies aries. Anim. Behav. 13, 249-258.

PERRY, G.C., PATTERSON, R.C.S. & STINSON, G.C. 1972. Submaxillary salivary gland involvement in porcine mating behaviour. VII. Int. Konf. Tierische Forpplanzunof. 1, 395-399.

PETERS, J.B., FIRST, N.L. & CASIDA, L.E. 1969. Effects of piglet removal and oxytocin injections on ovarian and pituitary changes in mammilectomised post partum sows. J. Anim. Sci. 28, 537-541.

PETROPAVLOVSKII, V.V. & RYKOVA, R.A. 1958. The stimulation of sexual functions in cows. Anim. Breed. Abstr. 1961, 29, No. 798.

POLIKARPOVA, E.F. 1960. Biological characters in livestock production. Anim. Breed. Abstr. 1961, 29, No. 1506.

POOLE, T.B. 1973. The aggressive behaviour of individual polecats (Mustele putorius M. furon and hybrids) towards familiar and unfamiliar opponents. J. Zool. 170, 395-414.

PORTER, A.R. Dairy cows vary in roughage preferences. Iowa Farm Sci. 7, 205-206.

PROCTOR, J., HOOD, A.E.M., FERGUSON, W.S. & LEWIS, A. H. 1950. The close-folding of dairy cows. J. Brit. Grassl. Soc. 5, 243-250.

PRYCHODKO, W. & LONG, A.P. 1969. Effect of isolation on the body weight of laboratory mice. Anatom. Rec. 138, 377.

RADFORD, H.M., WATSON, R.H. & WOOD, G.F. 1960. Crayon worn by ram to detect oestrus in service in sheep. Aust. Vet. J. 36, 57.

RASMUSSEN, O.G., BANKS, E.M., BERRY, T.H. & BECKER, D.T. 1962. Social dominance in gilts. J. Anim. Sci. 21, 519-522.

RATCLIFFE, D.A. 1965. Grazing in Scotland and upland England. In Grazing Experiments and the Use of Grazing as a Conservation Tool. Monks Wood Experimental Station, England. Symp. No. 2.

REED, H.C.B. 1969. Artificial insemination and fertility in the boar. Brit. Vet. J. 125, 272.

REINEKE, E.P., GARRISON, E.R. & TURNER, C.W. 1941. The relation of mastitis to the level of ascorbic acid and certain other constituents in milk. J. Dairy Sci. 24, 41-50.

ROBERTSON, I.S. 1966. Castration in farm animals: its advantages and disadvantages. Vet. Rec. 78, 130-140.

ROE, R. & MATTERSHEAD, B.E. 1962. Palatability of Phalaris arundinacea. L. Nature 193, 255-257.

ROGERS, C. D. Phil. Thesis Univ. Sussex.

ROSE, J.H. 1963. Ecological observations and laboratory experiments on free-living stages of Cooperia oncophora. J. Comp. Path. Ther. 73, 285-296.

ROSEN, J. 1958. Dominance behaviour as a function of post-weaning gentling in the albino rat. Cand. J. Psychol. 12, 229-234.

ROSENBLATT, J.S. & LEHRMAN, D.S. 1963. Maternal behaviour of laboratory rat. In Maternal Behaviour of Mammals. Ed. H.L. Rheingold. 8-55.

ROSS, O. 1960. Swine housing. Agric. Engin. 41, 584-585.

ROSS, S., ROSS, J.G. 1949. Social facilitation of feeding behaviour in dogs. I. Group & solitary feeding. J. Genetic. Psychol. 74, 97-108.

ROSS, S. & SCOTT, J.P. 1949. Relationship between dominance and control of movement in goats. J. Comp. Physiol. Psychol. 42, 75-80.

ROSS, S., SAWIN, P.B., ZARROW, M.X., & DENENBERG, V.H. 1963. Maternal behaviour in the rabbit. In Maternal Behaviour of Mammals. Ed. H.L. Rheingold, Wiley, 94-121.

ROSSDALE, P.D. 1968. Abnormal perinatal behaviour in the thoroughbred horse. Brit. Vet. J. 124, 540.

ROSSDALE, P.D. 1970. Perinatal behaviour in the thoroughbred horse. Brit. Vet. J. 126, 656.

ROWELL, T.E. 1966. Hierarchy in the organisation of a captive baboon group. Anim. Behav. 14, 430-443.

ROWELL, T.E. 1967. A quantitative comparison of the behaviour of wild and caged baboon groups. Anim. Behav. 15, 499-509.

ROWELL, T.E. 1969. Effect of social environment on menstrual cycles of baboons. Preliminary report. J. Reprod. Fert. Suppl. 6, 117-118.

SACKETT, G.P. 1965. Effects of rearing conditions upon monkeys (M. mullata). Child Development 36, 855-868.

SAHMARDANOV, Z.A., 1967. Changes in the thickness of the skin and wool covering of sheep up to and after removal of the gonads. Anim. Breed. Abstr. 36, 443.

SAMBRAUS, H.H. 1971. Das sexualverhalten des Hausrindes speciell des Stieres. Beinhaft 6, Zeit. fur Tierpsychol. Verglag Paul Parey, Berlin.

SANCTUARY, W.C. 1932. A study of avian behaviour to determine the nature and persistence of order of dominance in domestic fowl and to relate this to certain physiological reactions. M. Sc. Thesis, Mass. State College.

SASSENRATH, E.N. 1970. Increased adrenal responsiveness related to social stress in Rhesus monkeys. Hormones & Behav. 1, 283-298.

SCHAIBLE, P.J., DAVIDSON, J.A. & BANDERMER, S.L. 1947. Cannibalism and feather pecking in chicks as influenced by certain changes in specific ratios. Poultr. Sci. 26, 651-656.

SCHEIN, M.W. & FOHRMAN, M.H. 1954. Social dominance relationships in a herd of dairy cattle. Br. J. Anim. Behav. 1, 283-298.

SCHEIN, M.W. & HALE, E.B. 1959. The effect of early social experience on male sexual behaviour of androgen injected turkeys. Anim. Behav. 7, 189-200.

SCHEIN, M.W., HYDE, C.E. & FOHRMAN, M.H. 1955. The effect of psychological disturbances on milk production of dairy cattle. Proc. Ass. Southern Agric. Wkrs 52nd Convention, 79-80.

SCHINCKEL, P.G. 1954. The effect of the presence of the ram on the ovarian activity of the ewe. Aust. J. Agric. Res. 5, 465-469.

SCHIØRRING, E. & RANDRUP, A. 1968. 'Paradoxical' stereotyped activity of reserpinised rats. Int. J. Neuropharmcol. 7, 71-73.

SCHJELDERUP-EBBE, T. 1931. Die Despotie in sozialen leben der vogel. Forsch. Volkerpsychol. sozialog. 10, 77-140.

SCHMISSEUR, E.E., ALBRIGHT, J.L., DILLON, W.M., KEHRBERG, E.W. & MORRIS, W.H.M. 1966. Animal behaviour responses to loose and free stall housing. J. Dairy Sci. 49, 104-106.

SCOTT, E.M. 1946. Self selection of diet. 1. Selection of purified compounds. J. Nutr. 31, 397.

SCOTT, E.M. & QUINT, E. 1946. Self selection of diet. IV. Appetite for protein.

J. Nutr. 32, 293-300.

SCOTT, J.P. 1945. Social behaviour, organization and leadership in a small flock of domestic sheep. Comp. Psychol. Monograph 18, Ser. No. 96.

SCOTT, J.P. 1956. Analysis of social organization in animals. Ecology 37, No. 2.

SELYE, H. 1950. The physiology and pathology of exposure to stress. Acta Inc.

SEWARD, J.P. 1945. Aggressive behaviour in the rat. II. An attempt to establish a dominance hierarchy. J. Comp. Psychol. 38, 213-224.

SHELLEY, H.P. 1965. Eating behaviour: social facilitation or social inhibition. Psychonomic Sci. 3, 521-522.

SHORT, R.V. 1970. Social and sexual behaviour in red deer stags. Br. Vet. J. 120, 657.

SHRECK, P.K., STERRIT, G.H., SMITH, H.P., & STILSON, D.W. 1963. Environmental factors in the development of eating in chicks. Anim. Behav. 11, 306-309.

SHULTZE, J.V., JENSON, L.S., CARVER, J.S., & MATSON, W.E. 1960. Influence of various lighting regimes on the performance of growing chickens. Wash. Expt. St. Tech. Bull. 36, 1-11.

SIEGEL, H.S. 1959. The relation between crowding and weight of adrenal glands in chickens. Ecology 40, 495-498.

SIGNORET, J.P., BOUISSOU, M.F., & BUSNEL, N.R.G. 1960. Role d'un signal acoustique de verrat dans le comportment de la truie en oestrus. Comptes Rendus Ac. Sc 250, 1355.

SIGNORET, J.P. & DU MESNIL 1961. Etude du comportement de la truie en oestrus. Proc. 4 Int. Congr. Anim. Reprod. (The Hague) Physiol. Sec. 171-175.

SIREN, H.J. 1963a. A factor preventing cannibalism in cockerels. Life Sci. 2, 120-124.

SIREN, H.J. 1963b. Cannibalism in cockerels and pheasants. Act. Vet. Scand. 4, Supp. 1-48.

SKINNER, J.D. 1972. Sexual spermatogenesis in the black wildebeest, heartebeest and eland. Int. Conf. Reproduction and Fertility, Edinburgh.

SKOLUND, W.C. & PALMER, D.H. 1961. Light intensity studies with broilers. Poultr. Sci. 40, 1968.

SLEN, S.B. & CONNELL, R. 1958. Canad. J. Anim. Sci. 38, 38-47.

SMITH, W. 1957. Social 'learning' in domestic chicks. Behav. II, 40-55.

SOUTHCOLT, W.H. 1962. Austr. Vet. J. 38, 33.

SOUTHERLAND, G.F. 1939. Salivary conditioned reflexes in swine. Amer. J. Physiol. 126-640.

SOUTHWICK, C.H. 1966. Aggressive behaviour of the Rhesus monkey in natural and captive groups. In Aggressive Behaviour. Ed. Garattini and Sigg. 32-43. Excepta Medica.

SOUTHWICK, C.H. & BLAND, V.P. 1959. Effect of population density on adrenal glands and reproductive organs of CFW mice. Amer. J. Physiol. 197, 111-114.

STAPLETON, G.R. 1948. Pastures old and new. Agric. LV, 6.

STEIN, J.A.G., WINOKUR, A., EISENSTEIN, TAYLOR, R., & SLIG, M. 1960. The effects of group versus individual housing on behaviour and physiological responses to stress in the albino rat. J. Psychosomatic Res. 4, 185-190.

STEPHENS, D.M. & BALDWIN, B.A. 1970. Observations on the behaviour of artificially reared lambs. Br. Vet. J. 126, 659.

STEVENS, D.B. 1974. Studies on the effect of social environment on the behaviour and growth rates of artificially reared male calves. Anim. Proc. 18, 23-24.

SWANN REPORT 1969. On the use of antibiotics in agriculture. H.M.S.O.

SYME, G.H., SYME, L.A. & JEFFERSON, T.P. 1974. A note on the variations in the level of aggression within a herd of goats. Anim. Prod. 15, 309-312.

TEAGUE, H.S. & GRIEFE, S.P. 1961. Proc. of 1961 Ohio Swine Day 8.

THIESSEN, D.D. 1964. Population density and behaviour: a review of theoretical and physical contributions. Tex. Rep. Biol. Med. 22, 266-314.

THOMPSON, W.R., & SCHAFFER, T.J. 1961. Early environmental stimulation. In Functions of Varied Experience. Ed. D.W. Fiske & S.S.R. Maddi 81-105.

THURBER, S.W., DUNBAR, J.R. & SMITH, D.P. 1966. Calif. Agric. 20, 12-14.

TINDELL, R. & CRAIG, J.V. 1959. Effects of social competition on laying house performance in the chicken. Poultry Sci. 38. 95-105.

TOLMAN, G.W. 1965. Emotional behaviour and social facilitation of feeding in domestic chicks. Anim. Behav. 13, 493.

TOLMAN, G.W. 1968. The role of the companion in social facilitation of animal behaviour. In Social Facilitation and Imitative Behaviour. Ed. Simmiel, Hoppe and Milton. Algar and Barcon.

TOMKINS, T. & BRYANT, M.J. 1973. The influence of mature behaviour in the sheep on fertility at a progesterone-synchronized oestrus. Abstr. 56. Meeting Anim. Prod. 81.

TRIBE, D.E. 1950a. The behaviour of the grazing animal: a critical review of present knowledge. J. Brit. Grassl. Soc. 5, 209-224.

TRIBE, D.E. 1950b. The composition of the sheep's natural diet. J. Brit. Grassl. Soc. 5, 81-91.

TRIBE, D.E. 1950c. Influence of pregnancy and social facilitation on the behaviour of grazing sheep. Nature 166, 74.

TRIMBERGER, G.W. 1962. Artificial insemination. In Reproduction in Farm Animals. Ed. E.S. Hafez. Philad., Len & Fiebiger.

TUGAI, L.N. 1967. Skeletal growth of young black pied bulls on moderate feeding as affected by castration. Anim. Breed. Abstr. 36, 2346.

TURNER, C.D. 1961. General endocrinology. W.B. Saunders, London.

TURTON, J.D. 1962. The effect of castration on meat production and quality in cattle, sheep and pigs. Anim. Breed. Abstr. 30, 447.

TYLER, S. 1968. The behaviour of the New Forest ponies. Anim. Behav. Mon. 5.2.

ULRICH, R. 1966. Pain as a cause of aggression. Amer. 200, 6, 643-662.

VALENSTEIN, S.E. & YOUNG, W.C. 1965. Experimental and genetic factors in the organization of sexual behaviour in male guinea pigs. J. Comp. Physiol. Psychol. 48, 397-403.

VAN DEMARK, M.L. & HAYS, B.L. 1952. Uterine mobility responses to mating. Amer. J. Physiol. 170, 518-521.

VAN DER LEE-BOOT, C.M. 1956. Acta Physiol. Pharmacol. Neer. 5, 213.

VAN DER WELT, K. & JANSEN, B.C. 1968. Adaptation to stress and disease. In Adaptation of Domestic Animals. Ed. E.S.E. Hafez.

VAN LAWICK-GOODALL, J. 1968. The behaviour of free living chimpanzees in the Gombe Station Reserve. Anim. Behav. Monogr. 1. 3.

VAN LAWICK-GOODALL, H. & J.V. Innocent Killers. Collins 1970.

VAN PUTTEN, G. 1969. An investigation into tail biting among fattening pigs. Brit. Vet. J. 125-511.

VANDENBURG, J.G. 1971. The effects of gonadal hormones on the aggressive behaviour of adult golden hamsters. Anim. Behav. 19, 589-594.

VERSTEGEN, M.W.A. & VAN DER HEL 1974. The effects of temperature and type of floor on metabolic rate and effective critical temperature in groups of growing pigs. Anim. Prod. 18, 1-11.

VOGEL, H.H., SCOTT, J.P. & MARSDEN, M. 1950. Social facilitation and allelomimetic behaviour in dogs. Behav. 2, 121-143.

VON DER AHE, C. 1966. Anim. Breed. Abstr. 35, 620.

WAGNON, K.A. 1965. Social dominance in range cows and its effect on supplemental feeding. Div. Agr. Sci. Univ. Calif. Bull. 819.

WAGNON, K.A., LOY, R.G., ROLLINS, W.C. & CARROLL, R.D. 1966. Social dominance in a herd of Angus, Hereford and Shorthorn cows. Anim. Behav. 14, 474-479.

WALKER, D.M. 1950. Observations on behaviour of young calves. Bull. Anim. Behav. 8, 5-10.

WALKER, D.E. 1950. N.Z. J. Sc. Tech. 31, 30-38.

WALSTRA, P. & KROESKE, D. 1968. World Rec. Phys. Psycho. 49, 1-9.

WARNICK, A.C., CASIDA, L.E. & GRUMMER, R.H. 1950. The occurrence of oestrus and ovulation in post-partum sows. J. Anim. Sci. 9, 66.

WEININGER, O. 1956. The effects of early experience on behaviour and growth characteristics. J. Comp. Phys. Psych. 49, 1-9.

WEIR, W.C. & TORRELL, D.T. 1959. Selective grazing by sheep as shown by a comparison of the chemical composition of range and pasture forage obtained by handclipping and that collected by oesophageal fistulated sheep. Anim. Sci. 18, 641-649.

WEISMAN, U. 1956. Verhaltenstudien an der stockent (Anas platyrhynches L.). Das Aktiensystem. Zeit. fur Tierpsychol. 13, 208-271.

WELCH, R.H.S. & KILGOUR, R. 1970. Mis-mothering among Romneys. N.Z. Agric. 121, 26-27.

WELSH, B.L. 1964. Psychophysiological response to the mean level of environmental stimulation: a theory of environmental integration. In Medical Aspects of Stress in a Military Climate. U.S. Government Printing Office, Washington.

WESLEY, F. & KLOPFER, F.D. 1962. Visual discrimination learning in swine. Zeit. fur Tierpsych. 19, 93-104.

WHITTEN, W.K. 1956. Modification of mouse oestrous cycle by external stimuli associated with the male. J. Endocrin, 13, 399-404.

WHITTLESTONE, W.C., KILGOUR, R., DE LANGEN, H. & DUIRS, G. 1970. Behavioural stress and the cell count of bovine milk. J. Milk. Fd. Technol. 33, 217-220.

WIERZBOWSKI, S. 1964. Comparison of some characteristics of sexual behaviour of bulls, rams and stallions. 5th Cong. Int. Rep. Anim. & Artif. Fertil. 351-355.

WIGGINS, F.L., TERRIL, C.G. & EMIK, L.O. 1953. Relationships between libido and semen characteristics and fertility in range rams. J. Anim. Sci. 12, 684-696. 1970.

WILCOX, J.C. 1968. The effect of time and method of castration on performance of fat lambs. Expt. Husb. 17, 52-58.

WILKINSON, M. 1971. Seminar to Ethology Group, University of Sussex (in press).

WILLIS, E.N. 1966. Fighting in pigeons relative to available space. Psychom. Sci. 4, 315-316.

WILLNER, J.H., SAMACH, M., ANGRIST, B.M., WALLACE, B.B. & GERSHON, S S. 1970. Drug induced stereotyped behaviour and its antagonism in dogs. Com. Behav. 5, 135-141.

WILTBANK, J.N. & COOK, A.C. 1958. The comparative reproductive performance of nursed cows and milked cows. J. Anim. Sci. 17, 640-648.

WINGERT, F.L. & KNODT, C.B. 1960. Effects of the total floor space allocation for swine during the finishing period. J. Anim. Sci. 19, 1300.

WISMER-PEDERSEN, J. 1968. Boars as meat producers. World Rev. Anim. Prod. IV, 19-20 and 100-109.

WOOD, P.D.P., SMITH, G.F. & LISLE, M.F. 1967. A survey of intersucking in dairy herds in England and Wales. Vet. Rec. 81, 396-398.

WOODBURY, A.M. 1941. Changing the hook order in cows. Ecology 22, 410.

WOOD-GUSH, D. 1958. Fecundity and sexual receptivity in the brown Leghorn female. Poult. Sci. 37, 30-33.

WOOD-GUSH, D.G.M. 1971. The behaviour of the domestic fowl. Heinemann.

WOOD-GUSH, D.G.M. 1971. The Behaviour of the Domestic Fowl. Heinemann.

WOOD-GUSH, D.G.M. & KARE, M.R. 1966. The behaviour of calcium-deficient chickens. Br. Poult. Sci. 7, 285-290.

YAGEV, V. 1962. Effect of castration of male calves at 6 - 9 months of age on
    body conformation and health. Anim Breed. Abstr. 32, 88.
YEATES, N.T.M. & CROWLEY, T.A. 1961. Detection of oestrus in cattle. Past.
    Rev. & Graziers Rec. Vol. 71.
YEATES, N.T.M. 1965. Modern Aspects of Animal Production. Butterworths.
    London.

ZARROW, M.X. & DENENBERG, V.H. 1964. Maternal behaviour in the rabbit.
    Purdue Univ. Final Progress Report.
ZIEGENHAGEN, E.H. CARMAN, L.B. & HAYWARD, J.W. 1947. Feed particle
    size as a factor affecting performance of turkey poults. Poult. Sci. 26,
    212-214.
ZUCKERMAN, S. 1932. The Social Life of Monkeys and Apes. Kegan Paul,
    Trench, Trubner & Co.